An introduction to spectral analysis J.N.Rayner

Monographs in spatial and environmental systems analysis

Series editors R.J.Chorley and D.W.Harvey

p Pion Limited, 207 Brondesbury Park, London NW2 5JN

An introduction to spectral analysis J.N.Rayner

Pion Limited, 207 Brondesbury Park, London NW2 5JN

© 1971 Pion Limited

Library edition   SBN 85086 026 1
Student edition   SBN 85086 027 X
LC. 72-177346

Set on IBM 72 Composers by Pion Limited, London
Printed in Great Britain by J.W.Arrowsmith Limited, Bristol.

Preface

Early in his graduate work in climatology at McGill University the author was introduced to, and became fascinated by, the application of periodic analysis to the spatial study of the general circulation of the atmosphere. At that time he obtained a copy of J. W. Tukey's (1949) paper on the statistical aspects of spectral analysis of non-periodic data. Upon reading this paper again two or three years later he became aware of its true significance in the analysis of meteorlogical time series and began to apply the technique himself. In 1966 the author became associated with a number of quantitative geographers, particularly L. J. King, who were interested in the analysis of spatial series in general. It was as a result of this interest that the author was asked to develop a one quarter course for graduate students in geography and others at the Ohio State University. It soon became apparent that most of the students lacked the mathematical background to read in detail the relevant papers in the field. Furthermore, unlike their counterparts in engineering they were not versed in the concepts and uses of frequency response. Consequently, introductory sessions were necessary on trigonometric series and complex numbers. At the other extreme very little explanatory material existed on the expansion of the basic ideas to two dimensional problems and even the engineering students, with the exception of those using optical techniques and pattern recognition procedures, required some guidance. The author's reasons for writing this text then are typical; he needed some core reading which was sufficiently simple, yet sufficiently broad in scope, to provide a base and framework for the students in their private reading.

Because the author's major interests lie in climatology some bias is apparent, particularly in the exercises toward meteorology and geology and associated areas. However, since many of his students have been interested in a variety of fields, especially the social and behavioral environment, the writer has tried to reduce discussion of the specialised meaning of the results to a minimum. This may have been a bad decision, for it may mislead the reader into thinking that the technique is more important than the interpretation and future use of the results. Nothing could be further from the truth. This text deals with a number of techniques such as spectral regression, filtering, pattern recognition and differentiation under one cover because they have Fourier analysis in common, which in turn facilitates understanding; knowledge and understanding of one procedure is quickly and easily translated to another. Consequently, the organisation and scope of the text is one of convenience and not one of philosphy. It is the writer's contention that whereas there is a basis for arguing that the descriptive statistics for a sequenced series should include the spectrum of the variance as well as the mean and variance, the selection of a particular procedure should be dictated by the problem and not by one's attachment to some specialised method.

The author is indebted to the old College of Commerce and Business Administration at Ohio State for a grant to study the optical means of two dimensional spectral analysis, and to G. Sande who provided him with his fast Fourier transformation computer subroutines. Special thanks are due to my wife who gave constant moral support and typed several versions of the manuscript.

J. N. Rayner
Department of Geography, Ohio State University

Contents

1

Introduction

1.1 Aims

This set of notes has been written for the researcher in fields such as geography who is interested in finding out more about spectral analysis: what it does and how it may be applied. Good books (Blackman and Tukey, 1958; Granger and Hatanaka, 1964; Jenkins and Watts, 1968) already exist on the subject but it is assumed that the reader finds these difficult to follow through lack of background. Here it is assumed only that the reader is familiar with the concept of statistical significance, statistical techniques such as multiple regression, and knows what differentiation and integration mean. No mathematical proofs are given but if required they may be found in the reference material. A large number of formulae are given in the text but they are developed step by step so the reader, new to the field, should be able to follow them and therefore have a deeper understanding of the results.

One problem a researcher frequently has in reading up on an area which is new to him is in moving from one paper to another on the same subject but which uses different symbols and subscripts. In these notes an attempt is made to keep the notation consistent throughout and to keep the number of symbols down to a minimum. Each symbol and form of subscript will be explained when it is first met in the text and a complete list is given in Appendix A.

1.2 What is a spectrum and what is spectral analysis?

To most people the term 'spectrum' brings to mind the separation of colors in white light. In seeing the different colors the eye, with the brain, is in effect recognizing the different wavelengths of the visible portion of electromagnetic radiation: each group of wavelengths creates a different color from violet at $3 \cdot 5 \times 10^{-5}$ cm to red at 7×10^{-5} cm. The word 'spectrum' may be applied to other wavelengths and has been extended to phenomena other than radiation. Here the term will be applied to the characteristics of any phenomenon when those characteristics are ordered according to increasing or decreasing wavelength.

A wave may be visualized in an ideal form as a sinusoidal curve (Figure 1.2.1) with the wavelength being the distance between similar

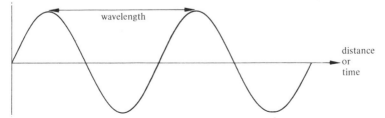

Figure 1.2.1.

points on the curve (e.g. from crest to crest). Where the horizontal axis is time, the words 'period' or 'cycle' are used in place of wavelength. A closely associated idea is that of frequency which is the number of cycles (waves) in a specified interval:

$$\text{frequency (in cycles per time unit)} = \frac{1}{\text{period}} , \qquad (1.2.1)$$

where the units are now often labelled 'Hz' (one hertz equals one cycle per second), or

$$\text{frequency (in cycles per distance unit)} = \frac{1}{\text{wavelength}} . \qquad (1.2.2)$$

In the particular case of a spatial wave moving at a constant average speed the two equations (1.2.1) and (1.2.2) are related:

$$\text{frequency (in cycles per time unit)} = \frac{\text{average speed}}{\text{wavelength}} . \qquad (1.2.3)$$

Therefore in the example of radiation, which travels at 3×10^{10} cm s^{-1}, red light has a frequency of approximately $4 \cdot 3 \times 10^{14}$ c/s.

It should be clear that the term frequency has a slightly different meaning in everyday language. For example, the frequency of vehicles using a tollgate on a freeway is usually thought of as the total number passing in a particular time interval. The order and separation time in which they pass is of no importance. On the other hand, in the spectrum, order and separation are just as important as the total number, and interest is centered upon the interval between occurrences or groups of occurrences. For instance, at the tollgate large numbers of vehicles may pass between 7 a.m. and 9 a.m. as compared with the rest of the day and more may be recorded during the middle of the week than at weekends. Such spacings of occurrences may be characterized by their frequencies of one cycle per day (seven cycles per week) and one cycle per week respectively. Other frequencies will be present and together they may be used as index variables for classifying the characteristics of the phenomenon. In the present example it might be of interest to know how the tollgate usage varied with increasing or decreasing frequency. The resulting distribution would be called the spectrum of tollgate usage.

Put in another way the spectrum may be considered as the scale breakdown of a phenomenon in space or time. In the tollgate example two scales, one diurnal and the other weekly, were introduced. Some scales, such as the latter, may be obvious from an examination of the raw data but others may be masked. Moreover, it may not be so evident how the scales are ranked, say in terms of variance. These features can only be brought out explicitly by a spectrum.

The calculation of a spectrum involves the fitting by least squares of sinusoidal curves of different frequencies to a set of data which may be in one, two, or n dimensions. Thus the method is equivalent to multiple regression with trigonometric transformations of the independent variable. However, the fitting turns out to be relatively simple because the functions are orthogonal and the parameters (coefficients of regression) are written in terms of simple sums of products. The number of coefficients obtained is equal to the number of datum points and the process known as a Fourier transformation. A return to the original observations from the coefficients involves another Fourier transformation. In each direction the data or the Fourier coefficients must be ordered or sequenced according to the units of space and/or time, or to the units of frequency or wavelength. The interval between the observations should be uniform but this is not necessarily over-restrictive since irregularly spaced observations may be fitted into a fine regular grid and simplifying assumptions made about the intervening grid points. For instance, Tobler (1969c) has used a number of different theoretical curves to spread populations over points adjacent to the centers of settlements. Where this is not feasible methods for using irregularly spaced data themselves are available (Neidell, 1967).

Spectral analysis in the present context refers primarily to the process of calculating and interpreting a spectrum. As such the aim is to seek deeper understanding of data and of the systems which produced those data. Depending upon whether the data are assumed to be periodic or not the resulting spectrum may take on two different forms. In the former, where the data repeat themselves indefinitely both forwards from the end and backwards from the beginning, as with atmospheric pressure around a latitude circle, the spectrum will be a discrete distribution made up of a finite number of frequencies (i.e. coefficients from regression). In the latter, and more usual non-periodic case, the frequencies are assumed to be continuous with each coefficient representing a band of frequencies which combine to produce a continuous density spectrum (Figure 1.2.2). An analysis of tollgate usage would produce a continuous spectrum. In essence the arithmetic involved in both forms is very similar but the results and their statistical confidence are very different.

Also spectral analysis may be stretched to include those instances where the spectrum is used as a simplifying step in a set of calculations. For example, some forms of pattern recognition are performed more easily

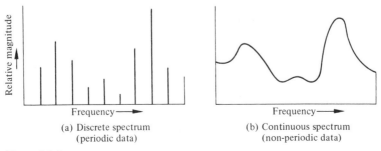

(a) Discrete spectrum
(periodic data)

(b) Continuous spectrum
(non-periodic data)

Figure 1.2.2.

with the spectrum than with the raw data. In fact, manipulations are performed on the Fourier coefficients themselves so that the discrete spectrum is the one used in this instance.

With this short introduction to what spectral analysis is, the remainder of this chapter will be devoted to a brief historical review of the development of the technique and a summary of its range of application.

1.3 Historical development

1.3.1 The discrete spectrum

It is generally accepted that Pythagoras, who was active about 530 B.C., was the first to recognize the importance of harmonics in music and it is well known that Kepler in the 16th and 17th centuries extended the idea of harmony to the motion of the planets (*Encyclopaedia Britannica*, 1965, see Harmonic Analysis). The first mathematical treatment of wave motion was given by Newton in his *Principia* in 1687 and the first publication of the equations, now known as the Fourier series, was made in 1728 by Daniel Bernoulli who worked with Euler on vibrating strings. Lagrange also discussed the technique in 1772. However, it was the work of Jean Baptiste Joseph Fourier which showed that almost any function of a real variable could be represented as the sum of sines and cosines (*Encyclopaedia Britannica*, 1965, see Fourier). The first announcement of this theory came in 1807 and it was stated fully in 1822 (Fourier, 1822). Since that time it has been put on a broader base with rigorous proofs by Dirichlet, Lebesgue, and Fejér. More recently the theory, which is the basis of all spectral analyses, has come to bear the name of Fourier rather than the adjective 'harmonic'.

Whereas Fourier also showed, by the Fourier integral, that periodicity is not a necessity for analysis, most applied researchers have restricted themselves to the periodic case and the discrete spectrum. This is unfortunate since usually their data are non-periodic. Early examples are the analyses of soil temperatures by Forbes (1846), Thomson (1860), and Everett (1860); and the analysis of tidal observations by Darwin (1883).

In 1897 the technique received wider publicity through Schuster (1897) who introduced a measure of statistical confidence into his analysis of the lunar and solar periodicities of earthquakes (see Section 3.5). The following year (Schuster, 1898) coined the term 'periodogram' for his spectrum and used it in a series of papers on the search for hidden periodicities in meteorological and astronomical data (Schuster, 1898, 1900, 1906a, 1906b). Thereafter the method was used in a variety of fields. For example, Brownlee (1917) investigated the periodicity of measles epidemics in London from 1703 to 1917 and Beveridge (1921, 1922) attempted to isolate cycles in European wheat yields and prices. Recently, there has been a revival due to Horn and Bryson (1960) who used it in the analysis of mean monthly rainfall (Sabbagh and Bryson, 1962; Fitzpatrick, 1964; Rayner, 1965). In each of the above cases the data have been non-periodic and consequently in many instances the results have been unreliable and misleading (e.g. Kendall, 1945, chapter 4). The original significance test of Schuster has been modified through the years (Section 3.5) and it is now clear that the distribution of the ordinate in the discrete spectrum is proportional to χ^2 with 2 degrees of freedom. Consequently, the standard error of the estimate is of the same order of magnitude as the estimate itself.

It is obvious, then, that this form of analysis should be restricted to legitimate data such as to variables around a latitude circle (Wiin-Nielsen *et al.*, 1963; Peixoto *et al.*, 1964; also see Appendix B) and to other truly periodic phenomena (are there any?).

1.3.2 The continuous spectrum

The development of spectral analysis of non-periodic data is more closely linked to the statistical analysis of stochastic processes and will be viewed therefore from that angle.

The initial step in any analysis is usually the description of the phenomenon. Statistically this involves the calculation of the mean and the variance. When observations are independent and normal, these two statistics describe the distribution completely, but with sequenced data the observations are not usually independent. In other words, adjacent observations tend to be correlated. Consequently the degree of this correlation needs to be specified. It has long been recognised that the obvious method is to calculate the autocorrelation or autocovariance function (Jenkins, 1961, p.135; Bendat and Piersol, 1966, p.70). The mathematical background to the practical application of the autocorrelation function in diffusion studies was set down by Taylor (1920, 1938) although the function had been used previously, for example, by Clayton (1917) in the analysis of Mt. Wilson solar radiation data. Another example of its use is that of Seiwell (1949) but the literature is vast (see Wold, 1965). More recently the autocovariance function for two dimensions has been calculated by Katz and Doyle (1964) in the analysis of cloud photographs and by Kovásznay and Arman (1957) in a more general approach for the analysis of test dot patterns. Unfortunately, the correlogram or some similar distribution is difficult to interpret beyond the first few lags. Furthermore, it is a difficult problem to specify the statistical significance of the estimates. It was Wiener's work in 1930 (Wiener, 1930) which showed that the variance spectrum (i.e. the breakdown of the variance with frequency or scale) and the autocovariance function contained the same information about the data. Also, it was subsequently shown by Tukey (1949) and independently by Bartlett (1950) that the variance spectrum has a more stable distribution and that statistical confidence bands may be easily calculated. Another very important point argued by Tukey (1949, 1967) is that the variance spectrum is more easily interpreted than the autocovariance function.

Incidentally, the 1967 paper by Tukey contains a short explanation of why the word spectrum is often used as an adjective rather than the word spectral. The adjective spectral may refer to spectre, whereas spectrum cannot. Bogert and Parzen (1967) also make comments concerning the use of the term *the* spectrum, which suggests uniqueness. They make the point that there are many spectra all different from one another.

Since the exposition of the technique by Tukey and Bartlett and the widespread introduction of the digital computer, spectral analysis has been utilized in virtually every field of research. As pointed out by Fourier, harmonic analysis is not restricted to one dimension and neither is the continuous form of analysis (Wiener, 1930, p.172). Furthermore, it was shown by Wiener (1930, p.182) that the analysis of more than one variable at a time is possible. Consequently, as the computational power of digital computers has increased, so the range of problems, to which spectral analysis has been applied, has expanded. The result is that papers concerned with its mathematical and statistical niceties and its application abound.

One of the earliest applications of one-dimensional, one-, and two-variable spectral analysis was in small-scale turbulence in meteorology (Panofsky, 1953; and other authors) where it has had enormous success over the last decade. This work has been admirably summarized by Lumley and Panofsky (1964). Subsequently it was extended to larger-scale problems and to climatology (Rayner, 1967b). One of the reasons for its success has been the fact that many of the parameters of the analyses have a physical as well as a solely statistical interpretation. For example, the spectrum of temperature in the atmosphere is proportional to the scale breakdown of potential energy.

Other applications include acoustics (Noll, 1964), astronomy (Uberoi, 1955), economics (Granger, 1966), engineering (Press and Tukey, 1956), geology (Preston, 1966), geomorphology (Speight, 1965), oceanography (Munk and Snodgrass, 1957), and tree growth (Bryson and Dutton, 1961).

Two-dimensional one-variable spectral analysis was first performed digitally by a group at New York University on ocean wave data (Pierson, 1960) and this was followed by research on cloud patterns (Leese and Epstein, 1963). Paralleling the development of digital techniques, spectral analysis in two dimensions by way of optical methods expanded following the introduction of the laser with its coherent light (O'Neill, 1956; Cutrona *et al.*, 1960; Dobrin *et al.*, 1965; Goodman, 1968), although the basic principles were set down by Abbe in 1873. Numerous filters have been used to remove and to emphasize particular features and to measure areas. For example, Dobrin *et al.* (1965) demonstrated how certain interference patterns could be removed from seismic records through optical filtering, and Pincus and Dobrin (1966) showed how similar techniques might be used for grain size analysis and for a comparison of contour patterns.

With a suitably equipped laboratory such optical procedures are quickly and economically executed and apparently far exceed the capacity of digital techniques. On the other hand, the recent introduction of the fast Fourier transform has made possible rapid and large-scale digital computation which, furthermore, has the advantage of not being limited to two dimensions (Gentleman and Sande, 1966; Tukey, 1967; I.E.E.E., 1967). It is interesting to note that, although the fast Fourier transform algorithm is considered new, it had several very similar predecessors dating back to Runge (1903). For some strange reason the earlier versions never gained wide publicity or use (Cooley *et al.*, 1967a).

1.4 Areas of application

From the above short account it is clear that spectral analysis has wide and expanding application in a variety of fields. In order to emphasize and clarify this variety the following two subsections attempt to classify some of the different problems to which the spectrum or Fourier transform has been applied.

1.4.1 The spectrum as the object of analysis

Here the center of interest is the spectrum itself. The object is to obtain the best frequency analysis possible. As will become clear in the later chapters a single sample sequence can produce many different spectra depending upon the various filters and windows used in the calculations. Consequently the researcher should experiment with these various modifications and not be satisfied with a single analysis. It should be emphasized also that seldom will the resulting spectra be the end of a research problem. A spectrum provides a different and often extremely revealing way of looking at a sample sequence. The same information is contained in the original data as the spectrum but it is arranged differently. Therefore the analysis should be considered as one of a number of possible alternative steps in the process of scientific investigation. To be useful, spectral analysis should provide not necessarily concrete results but should suggest new lines of inquiry towards the solution of a problem.

As already stated, information which the spectrum may provide is essentially related to scale analysis. For instance, the variance spectrum of a single variable in one or two dimensions describes the relative scales at which variation takes place within the original sequence or array. A fairly uniform distribution (within statistical sampling bands) of variance with scale may indicate that the data were generated by a random process. On the other hand, spectra with significant fluctuations suggest that there may be separable generating processes. The next step is a search for physical, social, behavioral, or economic, etc., processes which might have produced such scales. An example may be cited from micrometeorology where, in the analysis of some short-term wind fluctuations, variance spectra tended to show one or two peaks at about one-third and/or three

cycles per minute. Their presence prompted further investigation and led
to the recognition that these peaks were produced by the two forms of
turbulence, thermal and mechanical respectively (Panofsky and McCormick,
1954).

In two dimensions, besides scale of variability, additional information,
in the form of orientation or alignment, is given by the spectrum. For
example, Pierson (1960) and N.A.S. (1963) have shown that the two-
dimensional spectrum of ocean waves may be interpreted in terms of the
proportion of energy in each scale group of waves and the direction in
which that energy is moving. Such information is required for forecasting
and explaining the generation of waves.

Something to be discussed in the next subsection is the use of spectral
techniques in the filtering of data. In research, modification or filtering
of the data frequently takes place although not always intentionally. For
instance, the way in which the variable is recorded often modifies the data.
The daily summation of births or of rainfall is one such form. Another is
the calculation of a running mean or an areal aggregation. The effect of
these processes on the data is scale-dependent: one which may be made
explicit through the calculation of the spectrum of the process (e.g.
Holloway, 1958). It may well turn out that a scale, which is thought to
be removed, remains but with an altered time or spatial origin. Such an
occurrence, if undetected, may completely nullify any conclusions drawn
from the modified data. Thus an attempt should be made to estimate the
frequency response of all filters. In reverse, if a certain group of scales
need to be removed, or emphasized, a filter may be designed to have that
particular effect.

Whereas the variance spectrum alone may provide significant information
in research using sequenced data, usually interest is directed towards
relationships between variables. Just as the spectrum gives the scale
breakdown of the variance, the cross spectra give the scale breakdown of
the regression parameters such as covariance and correlation coefficient.
Frequently variables are related at one scale (correlation approaching $+1$)
but not at another (correlation approaching 0). Simple regression would
tend to cancel out these relationships and give some average and perhaps
misleading correlation. One of the more useful cross spectra is the phase
spectrum which gives the temporal or spatial lag between two variables in
each frequency. For example, if a time series at place A were assumed to
be related to another series at place B, the phase spectrum would give the
lag time for a disturbance of a particular size to travel from A to B or
vice versa (Rayner, 1967b).

Somewhat analogous to the study of the effect of filters mentioned
above is the problem of how a variable responds to some modification
where the modifier cannot be directly specified. Here the input may be
correlated spectrally with the output in an attempt to obtain the frequency
response of the process. This has particular applications in engineering

where a fluctuating input may be amplified to a dangerous level during the process. For example, airplanes should be constructed so that their structure will not amplify the turbulence encountered in the atmosphere. Otherwise the plane would disintegrate (Press and Tukey, 1956; Jenkins, 1962, 1963; Jenkins and Watts, 1968). Similar examples may be suggested in the human environment which is continually subjected to impulses (inputs) and which is monitored by various surveys (outputs). However, little research has been conducted using spectral techniques, since the data series tend to be short and heterogeneous.

1.4.2 The spectrum as an aid to computation

As almost identical calculations as those which produce a discrete spectrum from the original data will reverse the process, the same computational subroutine may be utilized in either direction. Now, as it will be shown in the later chapters that some manipulations are performed more easily in the spectral domain than in the data domain, it seems an obvious choice to use the Fourier transformation in the same way as logarithms. Furthermore, this choice is considerably enhanced for two-dimensional variables when optical methods are used. Each transformation involves one lens, so conversions to the spectrum and back again may be performed at the same time on one optical bench.

The spectrum used in this way may be applied to interpolation, filtering, differentiation, integration, and pattern recognition. For the newcomer to these areas it should be noted that in each case there are techniques other than the spectral one of performing these calculations, and that the method chosen should be dictated by the specific problem at hand.

Interpolation involves the calculation of points between the original observations. Usually a polynomial is fitted to the original data by least squares and then used to compute the required points. In principle the use of the spectrum is very similar, since trigonometric functions are fitted by least squares. In practice the spectrum is calculated, a number of zeros, dependent upon the spacing of the required points, is added and the inverse transform performed. According to Lanczos (1956) the results may be in some cases more meaningful than those from polynomials, although close to the boundaries of the data they may be extremely unreliable.

Filtering encompasses a wide range of applications and may be stretched to include differentiation and pattern recognition. In the data domain, filtering is usually thought of as multiplying by a weighting function and summing over a group of terms. Calculation of the running mean is perhaps the simplest example. Generally is requires a large number of multiplications and additions for each final term. In the spectral domain, on the other hand, it requires only a single multiplication and no addition for each term. Trends and trend surfaces, which include only the low frequencies, are easily isolated by discarding all the higher frequencies

(multiplication by zero) before retransformation. Trend surfaces have
been extensively applied to geology to describe regional subsurface
topography and to indicate, through maps of residuals from trend, the
possible location of mineral deposits (Krumbein and Graybill, 1965,
chapter 13). In some cases orientation filters have been used to remove
and emphasize certain alignments. By this technique Bauer *et al.* (1967)
were able to discover crevasses in the Greenland ice cap from an aerial
photograph.

One particular type of filter is the $-1, +1$ finite differencing operator.
Closely related, although often quite different in result, is differentiation.
Since the spectrum essentially represents a fitted mathematical function
to the original data, it may be differentiated. Practically this turns out
to involve a simple multiplication of the spectrum by frequency. The
inverse transform produces the first derivative of the original data. Rates
of change of variables are of universal interest since they are basic to the
study of dynamic systems. Also, in pattern recognition they may be used
to define and to enhance boundaries. Higher-order derivatives are
likewise easily obtained. Consequently, the Fourier series approach may
significantly simplify the calculations involved in the application of partial
differential equations. For example, by this method under certain
controlling assumptions, the vertical second partial derivative of a variable
may be obtained from discrete observations of that variable obtained in
the horizontal plane. (For applications to gravity and magnetic potentials
see Henderson and Zietz, 1949, and Darby and Davies, 1967.)

An expanding area of research at the present time is pattern
recognition. The research extends from medicine, where doctors want to
recognize automatically particular cells in microscope slides, to the Post
Office, where addresses need to be sorted. The problem has become acute
in the field of remote sensing where images from airplanes and satellites
are being obtained at a far greater rate than they can be analyzed.

One method proceeds by moving a known pattern over a large unknown
pattern and the calculation of the correlation coefficient. Areas of high
correlation will be similar to the known pattern. Analysis of this process
shows it to be no more than a filtering routine which is most readily
performed by a simple multiplication of the spectra of the known and
unknown patterns followed by inverse transformation. It should be
emphasized that this technique is really only a step in the recognition
process and that further processing usually will be necessary. However,
the technique is simple and fast and well adapted to the optical approach
(Goldstein and Rosenfeld, 1964; Tippett *et al.*, 1965; Cheng *et al.* (1968).

1.5 Remarks
With these applications in mind the step-by-step development of the
technique will now proceed. For a general outline the reader is referred
to the table of contents. In some areas a large number of references exist

but only the most pertinent ones appear in the text. For a more complete list, reference should be made to Appendix B.

Spectral analysis of non-periodic functions is introduced through a discussion of the 'classical' technique of transforming the modified autocovariance function, although the recent development of the fast Fourier transform algorithm (Gentleman and Sande, 1966) has made this technique obsolete. Nevertheless, many of the old problems carry over to the new method and at the present time much of the literature pertaining to them is in terms of the old method.

Discussion of the optical approach will be limited to general remarks since the results are similar to those obtained from digital analysis. The optical method at present has clear advantages in problems where the original data are already in picture form and in the relative ease with which the operator may modify filters in response to an observed output. On the other hand, there may be problems introduced by the specific optics of the particular system used and in the generation of quantitative hard copy results. Furthermore, with the development of picture input–output units for digital computers many of the advantages of the optical technique are lost.

1.6 Exercises

1.6.1 Plot on graph paper the following data using an abscissa (j) interval of $\frac{3}{4}$ inch and an $x[j]$ scale of $20°F = 1$ inch.

j	$x[j]$ (°F)	j	$x[j]$ (°F)
0	37	6	13
1	18	7	24
2	0	8	−8
3	−10	9	−21
4	−27	10	4
5	−10	11	35

Draw a continuous line joining all points. This line may be used for interpolation purposes and will be used for comparative purposes later.

1.6.2 List all the types of data which you use and which might be subject to a Fourier transformation.

2

Fourier series

2.1 Introduction to trigonometric functions

The trigonometric functions may be produced by the rotation of a vector of length V around the origin of a Cartesian grid. The angle measured anticlockwise from the abscissa to the vector may be given in degrees or radians, although the latter is more fundamental and will be used in the following. One radian is defined as the angle at the center of a circle subtended by an arc whose length is equal to the radius of that circle (Figure 2.1.1).

In all, 2π radii may be fitted around the circumference of a circle. Consequently in one revolution the vector passes through an angle of 2π radians, where $\pi = 3\cdot141\,592\,653\,589\,793\,238\,46\ldots$. Thereafter the vector passes through another cycle which is indistinguishable from the first, e.g. the vector at $5\pi/2$ is indistinguishable from the vector of the same magnitude at $\pi/2$. Therefore, if a trigonometric function is known completely for one revolution 2π, hereafter known as the *basic interval*, it is known for all revolutions or intervals. One complete revolution is equal to the arbitrary measure of $360°$. Thus one radian equals $57\cdot296°$.

From Figure 2.1.2 the three most frequently used trigonometric functions may now be defined by

$$\text{cosine}\,\theta = \frac{X}{V}\,,\tag{2.1.1}$$

$$\text{sine}\,\theta = \frac{Y}{V}\,,\tag{2.1.2}$$

and

$$\text{tangent}\,\theta = \frac{Y}{X} = \frac{\sin\theta}{\cos\theta}\,.\tag{2.1.3}$$

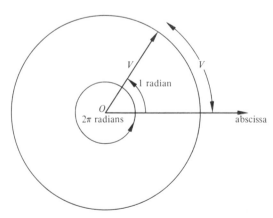

Figure 2.1.1.

These take on the sign of the ratios. For example, in the third quadrant X and Y are negative and V is positive. Therefore sine and cosine are negative and tangent is positive. A negative angle refers to one measured in the clockwise direction. For example,

$$\tan(-\tfrac{1}{3}\pi) = \tan\tfrac{5}{3}\pi$$

or

$$\tan(-60°) = \tan 300°.$$

Since, except for the signs, the ratios are similar in each quadrant, tables relating the angles to the ratios are given for the first quadrant only (Figure 2.1.3).

Let

$$\theta_2 = \pi - \theta_1, \qquad\qquad\qquad \theta_3 = \pi + \theta_1,$$

$$\theta_4 = 2\pi - \theta_1.$$

Then

$$\sin\theta_1 = \frac{Y}{V},$$

$$\sin\theta_2 = \sin(\pi - \theta_1) = \sin\theta_1 = \frac{Y}{V},$$

$$\sin\theta_3 = \sin(\pi + \theta_1) = -\sin\theta_1 = -\frac{Y}{V},$$

$$\sin\theta_4 = \sin(2\pi - \theta_1) = \sin(-\theta_1) = -\sin\theta_1 = -\frac{Y}{V}.$$

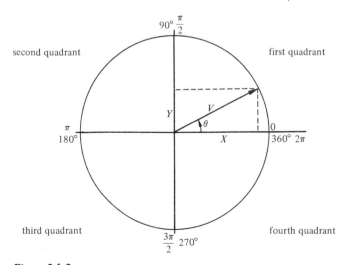

Figure 2.1.2.

Similarly, the cosines and tangents of angles in the second, third, and fourth quadrants may be related to those of the first quadrant. It will be noted that, whereas radians will be used in the text, tables are usually given in degrees. Computers, on the other hand, use radians (see Section 4.2.3).

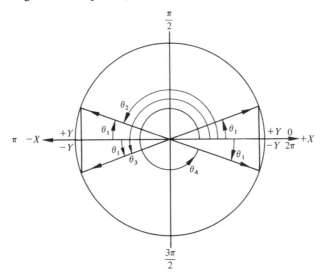

Figure 2.1.3.

2.2 The cosine curve

2.2.1 The basic curve

If V is taken to be of unit length ($V = 1$) and the magnitudes obtained from the cosine ratio are plotted against θ on a Cartesian grid, the result is as in Figure 2.2.1.1.

The equation of the curve is (θ in radians)

$$Y = \cos\theta \qquad 0 \leqslant \theta \leqslant 2\pi. \qquad\qquad (2.2.1.1)$$

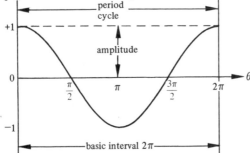

Figure 2.2.1.1.

2.2.2 Amplitude

The amplitude of the curve in Figure 2.2.1.1 is 1 but this may be altered by the multiplication of $\cos\theta$ by a factor A which may be any number, e.g.

$$Y = A\cos\theta .\qquad(2.2.2.1)$$

2.2.3 Frequency and period

One wavelength, one period, or one cycle is the distance between similar positions on the curve, such as from crest to crest. Frequency is the number of wavelengths (waves), periods, or cycles in some interval such as the basic interval.

The number of waves in the basic interval may be altered by the multiplication of θ by a factor k, which must be an integer, e.g.

$$Y = \cos(k\theta) .\qquad(2.2.3.1)$$

An example with $k = 2$ is drawn in Figure 2.2.3.1. The frequency of the function is 2 cycles (or waves) per basic interval and the period is $\frac{1}{2}$ basic interval. Period is therefore the reciprocal of frequency.

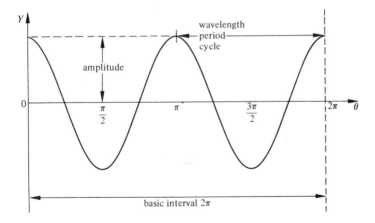

Figure 2.2.3.1.

2.2.4 Phase

The curve $Y = \cos(k\theta)$ starts with the crest at zero but it may start elsewhere if $k\theta$ is modified by the subtraction of a factor $\Phi[k]$, known as the phase angle, which may be any magnitude between 0 and 2π, e.g.

$$Y = \cos(k\theta - \Phi[k]) .\qquad(2.2.4.1)$$

Note that the phase angle is subscripted with k since it is associated specifically with the frequency k. Similarly, the amplitude associated with this frequency is subscripted in the same way.

From Equation (2.2.4.1), when $\theta = 0$, $Y = \cos(-\Phi[k])$. Also the curve reaches a maximum when $\cos(k\theta - \Phi[k]) = 1$ or when $k\theta - \Phi[k] = 0$, i.e. when $\theta = \Phi[k]/k$, which is known as the phase shift. An example is given in Figure 2.2.4.1 with $k = 2$, $\Phi[k] = \pi/3$.

Therefore, phase is the distance of the crest from the origin if it is assumed that a single wave (frequency, $k = 1$) is completed in the basic interval. The phase shift is the actual distance of the crest from the origin regardless of the frequency.

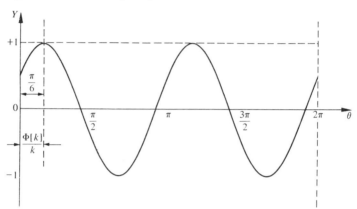

Figure 2.2.4.1.

2.3 Alternate representation

In general, then, a cosine curve with k waves per basic interval, amplitude $A[k]$, and phase angle $\Phi[k]$ may be written

$$Y = A[k] \cos(k\theta - \Phi[k]) \qquad 0 \leqslant \theta < 2\pi, \; 0 \leqslant \Phi < 2\pi . \qquad (2.3.1)$$

This may be expressed in a more useful and simple form as the sum of a sine and a cosine function. The alternate expression is obtained through the use of the trigonometric relationship

$$\cos(R - S) = \cos S \cos R + \sin S \sin R . \qquad (2.3.2)$$

Equation (2.3.1) may therefore be written

$$Y = A[k]\cos(k\theta - \Phi[k])$$
$$= A[k]\cos(\Phi[k])\cos(k\theta) + A[k]\sin(\Phi[k])\sin(k\theta) . \qquad (2.3.3)$$

Define $a[k]$ and $b[k]$ by

$$a[k] = A[k]\cos(\Phi[k]) \qquad (2.3.4)$$

and

$$b[k] = A[k]\sin(\Phi[k]) , \qquad (2.3.5)$$

then Equation (2.3.3) becomes

$$Y = A[k]\cos(k\theta - \Phi[k]) = a[k]\cos(k\theta) + b[k]\sin(k\theta) . \qquad (2.3.6)$$

$A[k]$ and $\Phi[k]$ may be obtained from $a[k]$ and $b[k]$. Since

$$\cos^2 S + \sin^2 S = 1 ,\tag{2.3.7}$$

the sum of the squares of Equations (2.3.4) and (2.3.5) gives

$$A[k] = (a^2[k] + b^2[k])^{\frac{1}{2}} .\tag{2.3.8}$$

Equation (2.3.5) divided by Equation (2.3.4) gives

$$\Phi[k] = \arctan\left(\frac{b[k]}{a[k]}\right)\tag{2.3.9}$$

where the arctangent (angle) of a ratio is obtained by searching the body of a tangent table for the ratio $b[k]/a[k]$ and reading off the corresponding angle. In other words, it is the reverse of looking up the tangent of the angle. The signs of $a[k]$ and $b[k]$ give the quadrant in which $\Phi[k]$ appears. For example, from Section 2.1, if both were negative,

$$\Phi[k] = \pi + \arctan\left(\frac{|b[k]|}{|a[k]|}\right) .$$

The same argument may be applied to the sine curve. Equation (2.3.6) then becomes

$$A[k]\sin(k\theta + \Phi[k]) = a[k]\cos(k\theta) + b[k]\sin(k\theta)\tag{2.3.10}$$

and the ratio in Equation (2.3.9) is inverted.

2.4 The sum of sinusoidal functions

The sum of periodic functions of the type given in Equations (2.3.6) is known as a Fourier series, i.e.

$$x[\theta] = \sum_{k=0}^{\infty} A[k]\cos(k\theta - \Phi[k])\tag{2.4.1}$$

$$= \sum_{k=0}^{\infty} \{a[k]\cos(k\theta) + b[k]\sin(k\theta)\}\tag{2.4.2}$$

$$= \bar{a}[0]\cos 0 + a[1]\cos\theta + ... + a[k]\cos(k\theta) + ...$$
$$+ b[0]\sin 0 + b[1]\sin\theta + ... + b[k]\sin(k\theta) + ...\tag{2.4.3}$$

Since $\sin 0 = 0$, there is no $b[0]$ term and, since $\cos 0 = 1$, $\bar{a}[0]$ stands alone and is equal to $A[0]$. $a[0]$ is made distinguishable from the other amplitudes because a slightly different equation is required for its calculation (see Section 3.3).

2.5 An example

Figure 2.5.1 shows an example with $k = 0, 1, 2$; $A = 6, 3, 2$; and $\Phi[k] = 0, \pi/4, 0$.

Thus it can be seen that the sum of simple waves can produce a complicated curve.

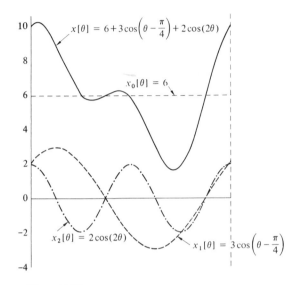

Figure 2.5.1.

2.6 Exercises

2.6.1 Write down the amplitudes ($A[k]$'s) and phases ($\Phi[k]$'s) in degrees corresponding to frequencies (k), 0 to 9, for the following functions:

(a) $x[\theta] = 7 + 2\cos\left(3\theta - \dfrac{\pi}{6}\right) + 5\cos\left(4\theta - \dfrac{\pi}{4}\right) + 6\cos\left(8\theta + \dfrac{\pi}{5}\right)$

(b) $x[\theta] = 4 + 3\cos(5\theta) + 5\cos(7\theta) + \sin(4\theta) + 4\sin(5\theta) + 12\sin(7\theta)$.

2.6.2 Draw curves for twelve points for $k = 0, 1, 3$; $A[k] = 2, 3, 4$; and $\Phi[k] = 0, \pi, \pi/2$. Sum and draw $x[\theta]$.

2.6.3 Find $a[k]$ and $b[k]$ for the curves in Exercise 2.6.2.

2.6.4 Reading: for basic trigonometry look at introductory texts such as Shapiro and Whitney (1967).

3

Fourier analysis

3.1 Basic assumptions

Harmonic or Fourier analysis is the process of fitting Fourier series to data and of calculating $A[k]$ and $\Phi[k]$, the amplitudes and phase angles, of the various waves. This is the reverse process of that shown in Figure 2.5.1. Thus a complicated function is reduced to a series of simple functions, sinusoidal waves. That this may be done for any function $x[t]$ may be proved mathematically provided that (1) $x[t]$ is single-valued and finite, (2) $x[t]$ is defined for every point in the basic interval, and (3) $x[t]$ has a finite number of maxima, minima, and discontinuities in the basic interval.

The fitting method used is that of least squares, but because of the orthogonality of trigonometric functions this technique is greatly simplified and the coefficients from the normal equations produce a diagonal matrix.

Since the function $x[t]$ is frequently represented by a series of discrete points (observations), the resulting Fourier series will depict the points and not necessarily the function $x[t]$. The closeness of fit between points and therefore the usefulness of the Fourier series for interpolation will depend upon the actual frequencies present in $x[t]$ and those calculable from the discrete points (see Section 6.4).

Under the above assumptions then, any sequence of data may be represented by a Fourier series. However, if any interpretation is to be made of the spectral results (amplitudes, phase angles, frequencies), some assumptions must be made about the function beyond the limits of the data. The simplest assumption and the one which is made for this section is that the function repeats itself completely to $-\infty$ and to ∞, i.e. that it is completely periodic. There are few truly periodic functions in nature, yet this assumption of periodicity leads to simple calculations which turn out to be basic to the alternative approaches to be dealt with in Chapters 5 and 7.

3.2 The independent variable

The function $x[t]$ has the independent variable t which may be in units of distance or time. For multi-dimensional situations, which will be dealt with later, $x[t_1, t_2]$ might represent a function which is dependent upon the coordinates t_1 kilometers east and t_2 kilometers south from some arbitrary origin. It is assumed that translation of the origin does not affect the calculations (see Section 7.1), so that $t = 0$ may be placed within the data. The phase angle, of course, must be adjusted.

The function will be observed over some interval of space (or time), T, which will be known as the basic interval. Seldom will this interval in the independent variable be measured in units of radians, so a conversion must be accomplished. So far θ has been the independent variable which varies between 0 and 2π. t varies between 0 and T. Therefore

$$\theta = \frac{2\pi t}{T} \text{ radians.} \qquad (3.2.1)$$

For continuous functions Equation (3.2.1) may now be used to replace θ in the previous equations.

If, as is usually the case, there are n equally spaced observations over the basic interval 2π, the spacing between successive observations is given by

$$\Delta\theta = \frac{2\pi}{n} \text{ radians.} \qquad (3.2.2)$$

Any particular observation may be denoted by a bracketed subscript, $[j]$. For example, the $(j+1)$th point along the abscissa may be written $\theta[j]$, where

$$\theta[j] = \theta[0] + j\Delta\theta = \theta[0] + j\frac{2\pi}{n} \qquad (3.2.3)$$

or, if $\theta[0] = 0$,

$$\theta[j] = \frac{2\pi j}{n} \text{ radians,} \qquad (3.2.4)$$

and where j is an integer varying between 0 and $n-1$. In other words, $\theta[0]$ will be the position of the first observation and $\theta[n-1]$ the last, with $\theta[n]$ $(= 2\pi)$ starting the next cycle. Thus Equation (3.2.4), which is not dependent upon the units of t, may be used to replace θ $(= \theta[j])$ in previous equations for equally spaced data.

On the other hand, if t is required, since

$$t = t[j] = j\Delta t, \qquad (3.2.5)$$

j may be replaced by $t/\Delta t$ in Equation (3.2.4) and so

$$\theta[j] = \frac{2\pi t}{n\Delta t} \text{ radians.} \qquad (3.2.6)$$

It will be seen that Equations (3.2.1) and (3.2.6) are equivalent and that $T = n\Delta t$. In the continuous case $\Delta t \to 0$ as $n \to \infty$.

The same relationship (3.2.6) may be used to convert the phase shift to the units of t, i.e.

$$\begin{matrix} \text{distance of the first maximum from} \\ \text{the origin in units of } t \end{matrix} = \frac{\Phi[k]}{k}\frac{n\Delta t}{2\pi}. \qquad (3.2.7)$$

The various notations can now be listed (see Table 3.2.1). As an example, Equation (2.4.1) will from this point on be written as

$$x[t] = \sum_{k=0}^{\infty} A[k]\cos\left(\frac{2\pi kt}{T} - \Phi[k]\right) \qquad (3.2.8)$$

for continuous data, or as

$$x[j] = \sum_{k=0}^{\infty} A[k]\cos\left(\frac{2\pi kj}{n} - \Phi[k]\right) \qquad (3.2.9)$$

for discrete equi-spaced data.

Since at least two points are needed to specify a sinusoidal curve, the maximum frequency k_{max} calculable from equi-spaced data is $n/2$, where n is even, or $(n-1)/2$, where n is odd. Consequently ∞ should be replaced by $n/2$ or $(n-1)/2$ in Equation (3.2.9). Calculation of coefficients for $k > k_{max}$ will show that they are periodic: the a's are symmetric and the b's asymmetric about $k = 0$ and $k = k_{max}$.

Table 3.2.1.

Sequential integer		0,	1,	...,	j,	...,	$n-1$
Independent variable	equi-spaced	0,	$\dfrac{2\pi}{n}$	...,	$\dfrac{2\pi j}{n}$,	...,	$\dfrac{2\pi(n-1)}{n}$
	continuous				θ		
					$\dfrac{2\pi t}{n\Delta t}$		
					$\dfrac{2\pi t}{T}$		
Dependent variable	equi-spaced	$x[0]$,	$x[1]$,	...,	$x[j]$,	...,	$x[n-1]$
	continuous				$x[\theta]$		
					$x[t]$		

3.3 Calculation of the coefficients

As mentioned above the coefficients $a[k]$ and $b[k]$ are easily found as $\cos(R\,2\pi t/T)$ and $\sin(S\,2\pi t/T)$ are orthogonal functions. Functions $x[t]$ and $y[t]$ are said to be orthogonal over the interval (u, v) if the integral of the product $x[t]y[t]$ over the interval vanishes:

$$\int_u^v x[t]y[t]dt = 0 \ . \qquad (3.3.1)$$

Such integrals for sine and cosine are stated below and will be used in simplifying later equations.

$$\left.\int_0^T \cos\left(R\,\frac{2\pi t}{T}\right)\cos\left(S\,\frac{2\pi t}{T}\right)dt \ \right\} = 0 \text{ for } R \neq S \qquad (3.3.2)$$

$$\left.\int_0^T \sin\left(R\,\frac{2\pi t}{T}\right)\sin\left(S\,\frac{2\pi t}{T}\right)dt \ \right\} = \frac{T}{2} \text{ when } R = S \ , \qquad (3.3.3)$$

$$\int_0^T \cos\left(R\frac{2\pi t}{T}\right)\sin\left(S\frac{2\pi t}{T}\right)dt \Bigg]$$ (3.3.4)

$$\int_0^T \sin\left(R\frac{2\pi t}{T}\right)dt \Bigg\} = 0 \text{ for all } R \text{ and } S ,$$ (3.3.5)

$$\int_0^T \cos\left(R\frac{2\pi t}{T}\right)dt \quad \begin{array}{l} = 0 \text{ for } R \neq 0 \\ = T \text{ when } R = 0 . \end{array}$$ (3.3.6)

Had θ been used, Equation (3.3.2) would read

$$\int_0^{2\pi} \cos(R\theta)\cos(S\theta)d\theta \quad \begin{array}{l} = 0 \text{ for } R \neq S \\ = \pi \text{ when } R = S . \end{array}$$ (3.3.7)

To obtain the coefficients $a[k]$ or $b[k]$ each side of Equation (2.4.3) must be multiplied by the corresponding trigonometric function, $\cos(k\,2\pi t/T)$ or $\sin(k\,2\pi t/T)$ and each side integrated over the basic interval T.

For example, to obtain $a[k]$, both sides of Equation (2.4.3) are multiplied by $\cos(k\,2\pi t/T)$, i.e.

$$\int_0^T x[t]\cos\left(k\frac{2\pi t}{T}\right)dt = \int_0^T \bar{a}[0]\cos\left(k\frac{2\pi t}{T}\right)dt + ...$$

$$+ \int_0^T a[1]\cos\left(\frac{2\pi t}{T}\right)\cos\left(k\frac{2\pi t}{T}\right)dt + ...$$

$$+ \int_0^T a[k]\cos\left(k\frac{2\pi t}{T}\right)\cos\left(k\frac{2\pi t}{T}\right)dt + ...$$

$$+ \int_0^T b[1]\sin\left(\frac{2\pi t}{T}\right)\cos\left(k\frac{2\pi t}{T}\right)dt + ...$$ (3.3.8)

Because of Equations (3.3.3), (3.3.4), (3.3.5), and (3.3.6) all terms except the one containing $a[k]$ on the right-hand side of (3.3.8) are zero. Equation (3.3.8) therefore reduces to

$$\int_0^T x[t]\cos\left(\frac{2\pi kt}{T}\right)dt = a[k]\frac{T}{2} .$$ (3.3.9)

Therefore

$$a[k] = \frac{2}{T}\int_0^T x[t]\cos\left(\frac{2\pi kt}{T}\right)dt$$ (3.3.10)

or, in summation terms,

$$a[k] = \frac{2}{n\Delta t}\sum_{j=0}^{n-1} x[j]\cos\left(\frac{2\pi jk}{n}\right)\Delta t .$$ (3.3.11)

Similarly, by multiplication by $\sin(2\pi kt/T)$ and integration

$$b[k] = \frac{2}{n}\sum_{j=0}^{n-1} x[j]\sin\left(\frac{2\pi jk}{n}\right)$$ (3.3.12)

and by integration alone

$$\bar{a}[0] = \frac{1}{n}\sum_{j=0}^{n-1} x[j] \qquad (3.3.13)$$

and for n even

$$\bar{a}\left[\frac{n}{2}\right] = \frac{1}{n}\sum_{j=0}^{n-1} x[j]\cos\left(\frac{2\pi j n/2}{n}\right) = \frac{1}{n}\sum_{j=0}^{n-1} x[j](-1)^j. \qquad (3.3.14)$$

There is no $b[n/2]$. It can be seen that $\bar{a}[0]$ is, in fact, the mean of the function $x[t]$. In order to make this coefficient comparable with the others the mean is frequently defined as being equal to $\frac{1}{2}a[0]$. Then

$$a[0] = \frac{2}{n}\sum_{j=0}^{n-1} x[j]\cos\left(\frac{2\pi j 0}{n}\right) \qquad (3.3.15)$$

which is the same as Equation (3.3.11) with $k = 0$. Similarly $\bar{a}[n/2]$ may be set equal to $\frac{1}{2}a[n/2]$ and

$$a\left[\frac{n}{2}\right] = \frac{2}{n}\sum_{j=0}^{n-1} x[j]\cos\left(\frac{2\pi j n/2}{n}\right) \qquad (3.3.16)$$

which is the same as Equation (3.3.11) with $k = n/2$. Therefore Equations (3.3.11) and (3.3.12) are sufficient for calculating all coefficients. $A[k]$ and $\Phi[k]$ are obtainable from Equations (2.3.8) and (2.3.9) with the exceptions of $A[0]$, which is $\bar{a}[0]$, and $A[n/2]$, which is $\bar{a}[n/2]$.

The amplitude estimates $A[k]$ may be plotted against frequency k to give a discrete line spectrum (Figure 3.3.1).

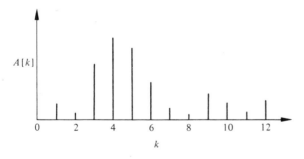

Figure 3.3.1. Discrete line spectrum.

3.4 Breakdown of the variance

It is evident that Fourier analysis isolates the separate harmonics present in the data. It is shown here that this is equivalent to componentizing the variance according to frequency. The importance of any frequency may then be gauged as its percentage contribution to the total variance.

The mean of the data is $\frac{1}{2}a[0]$, therefore

$$\hat{\sigma}^2_{total} = \sum_{j=0}^{n-1} \frac{(x[j]-\frac{1}{2}a[0])^2}{n} = \frac{\sum x^2[j]}{n} - \left(\frac{a[0]}{2}\right)^2$$

$$= \frac{1}{T}\int_0^T x^2[t]dt - \left(\frac{a[0]}{2}\right)^2 . \tag{3.4.1}$$

From Equation (2.4.3)

$$x^2[t] = \left(\frac{a[0]}{2} + a[1]\cos\frac{2\pi t}{T} + a[2]\cos\frac{2\pi 2t}{T} + ... + a[k]\cos\frac{2\pi kt}{T} + ...\right.$$

$$\left. + b[1]\sin\frac{2\pi t}{T} + b[2]\sin\frac{2\pi 2t}{T} + ... + b[k]\sin\frac{2\pi kt}{T} + ...\right)^2 .$$

Therefore

$$\frac{1}{T}\int_0^T x^2[t]dt = \frac{1}{T}\int_0^T \left(\frac{a[0]}{2} + a[1]\cos\frac{2\pi t}{T} + ... \text{etc.}\right)^2 dt ,$$

when the full expression for $x^2[t]$ shown above is substituted, but because of Equations (3.3.2) to (3.3.6) all terms on the right-hand side are zero except for those containing the squares of the coefficients, hence

$$\frac{1}{T}\int_0^T x^2[t]dt = \frac{1}{n}\left(\frac{a^2[0]}{4}n + a^2[1]\frac{n}{2} + a^2[2]\frac{n}{2} + ... + a^2[k]\frac{n}{2} + ...\right.$$

$$\left. + b^2[1]\frac{n}{2} + b^2[2]\frac{n}{2} + ... + b^2[k]\frac{n}{2} + ...\right)$$

and $A^2[k] = a^2[k] + b^2[k]$ with $A^2[n/2] = \bar{a}^2[n/2] = \frac{1}{4}a^2[n/2]$ for n even. Therefore

$$\hat{\sigma}^2[total] = \frac{A^2[1]}{2} + \frac{A^2[2]}{2} + ... + \frac{A^2[k]}{2} + ... \tag{3.4.2}$$

Note that the last term is $A^2[n/2]$ if n is even or $A^2[(n-1)/2]/2$ if n is odd.

Also the variance of each cosine curve is given by half the square of its amplitude, i.e.

$$\hat{\sigma}^2[k] = \frac{A^2[k]}{2} = \frac{a^2[k]+b^2[k]}{2} \tag{3.4.3}$$

with $\hat{\sigma}^2[n/2] = A^2[n/2]$ if n is even and the percentage contribution of any wave to the whole is given by

$$\text{contribution } (\%) = \frac{A^2[k] \times 100}{\sum_{k=1}^{n/2} A^2[k]} . \tag{3.4.4}$$

These variances of the individual frequencies may be plotted in the same way as the amplitudes to give the variance line spectrum. With percentages plotted the diagram shows the probability distribution of the frequencies (or wavelengths).

3.5 Significance of the estimates

The original data are usually only a sample of the universe and therefore the calculations give only estimates of the parameters. Consequently, the question now arises as to how reliable these estimates of the variance (or amplitudes) are at a given frequency.

Actually, there are two sampling problems. First, if the original function is continuous, as is usually the case, the taking of equi-spaced observations, which is a systematic sampling procedure, makes it impossible to isolate high frequencies. Instead these high-frequency variances will be added into the lower resolvable frequencies and may make them unusable. This is called aliasing and will be discussed in detail in Section 6.4. Secondly, the function $x[t]$ may be considered to be only one realization (one sample) of the process $x[t]$.

Generally in discrete analysis aliasing is ignored and the second more normal type of sampling problem considered alone. The usual approach, and the one taken by Schuster (1897, 1898) in the original discussions of the problem, is to assume that the raw data are independent and normally distributed. Then, using a theorem proved by Rayleigh (1880), Schuster was able to show that the estimate of the variance, standardized by the mean square, is distributed as an exponential.

Walker (1914) modified this procedure to take into account the fact that the largest rather than a randomly selected estimate was tested. He assumed that the population variance was known but Fisher (1929) removed that restriction.

Despite these tests harmonic (or periodogram) analysis produced misleading results (see, for example, Kendall, 1946). They prompted Kendall (1945, chapter 5, Section 2, p.434) to look more closely at the basic assumption of randomness in the original series and to suggest that it was not realistic. However, it should be noted that Schuster recognized this limitation (Schuster, 1898, Section 6, 1900, pp.124–127) as pointed out by Conrad and Pollack (1950, p.396).

In order to circumvent the problem of persistence (lack of independence of adjacent data in the series) Hartley (1949) introduced a new approach through the application of the analysis of variance and the use of the F test. Since it appeared at the time when Bartlett and Tukey introduced the concept of the continuous spectrum in a practical form, this has been little used.

At the present time, no satisfactory test of significance for the discrete spectrum of random data is available. It was early recognized that Schuster's distribution is equivalent to χ^2 with 2 degrees of freedom. Consequently, it is independent of n as shown, for example, by Jenkins (1961) and the standard error of the estimate is the same order of magnitude as the estimate.

3.6 Practical method of calculating the coefficients

3.6.1 The general method

At the present time the fastest method for calculating the Fourier coefficients is by the fast Fourier transform algorithm (Section 4.9). However, the full utility of that algorithm is best described through using complex numbers so an older less efficient method is used here.

The calculations of $a[k]$ and $b[k]$ may be simplified if Equations (3.3.11) and (3.3.12) are written out in full and use is made of the symmetry characteristics of sine and cosine functions.

For example, if $n = 12$ ($2\pi/n = 30°$), the equation for $a[1]$ becomes

$$a[1] = \tfrac{1}{6}(x[0]\cos\ \ 0+\ x[1]\cos\ \ 30+\ x[2]\cos\ \ 60+x[3]\cos\ \ 90$$

$$+x[6]\cos180+\ x[5]\cos150+\ x[4]\cos120$$

$$+\ x[7]\cos210+\ x[8]\cos240+x[9]\cos270$$

$$+x[11]\cos330+x[10]\cos300)\ .$$

This may be rewritten since $\cos30 = -\cos150 = -\cos210 = +\cos330$, etc.:

$$a[1] = \tfrac{1}{6}(x[0]\cos0+\ x[1]\cos30+\ x[2]\cos60+x[3]\cos90$$

$$-x[6]\cos0-\ x[5]\cos30-\ x[4]\cos60$$

$$-\ x[7]\cos30-\ x[8]\cos60-x[9]\cos90$$

$$+x[11]\cos30+x[10]\cos60)$$

$$= \tfrac{1}{6}\{(x[0]-x[6])\cos0+(x[1]-x[5]-x[7]+x[11])\cos30+$$

$$+(x[2]-x[4]-x[8]+x[10])\cos60+(x[3]-x[9])\cos90\}\ .$$

For $a[3]$ this equation is

$$a[3] = \tfrac{1}{6}\{(x[0]-x[6])\cos0+(x[1]-x[5]-x[7]+x[11])\cos90$$

$$+(x[2]-x[4]-x[8]+x[10])(-\cos0)+(x[3]-x[9])\cos90\}\ .$$

It will be seen that for odd cosines ($a[1], a[3], a[5]$, etc.) the same sums appear in the round brackets and that only the cosine multiplier changes. For even cosines ($a[2], a[4], a[6]$, etc.) different sums will appear in the round brackets. A similar arrangement may be verified for sine.

Since multiplication takes a much longer time than addition even on a computer, this method, collecting the sums of ordinate values together first, saves time. Furthermore, the following tabular method allows easy calculation on a desk calculator.

Step 1: draw up table of θ with each column representing a different quadrant and each row a different modulus value of sine or cosine.

Table 3.6.1.1.

$\theta = \dfrac{2\pi j}{n}$	Column 1	θ	Column 2	θ	Column 3	θ	Column 4
0		180					
30		150		210		330	
60		120		240		300	
90				270			

Step 2: beside θ put the corresponding value of x (i.e. $x[0], x[1], ..., x[11]$).

Step 3: draw up table for the sums of x for each row. Four sums may be obtained according to

odd sin I	column 1 + column 2 − column 3 − column 4
even sin II	column 1 − column 2 + column 3 − column 4
odd cos III	column 1 − column 2 − column 3 + column 4
even cos IV	column 1 + column 2 + column 3 + column 4

Table 3.6.1.2.

	column 1		column 2		column 3		column 4
0	$x[0]$	180	$x[6]$				
30	$x[1]$	150	$x[5]$	210	$x[7]$	330	$x[11]$
60	$x[2]$	120	$x[4]$	240	$x[8]$	300	$x[10]$
90	$x[3]$			270	$x[9]$		

Table 3.6.1.3.

	I	II	III	IV
0	A	E	I	M
30	B	F	J	N
60	C	G	K	O
90	D	H	L	P

Step 4: draw up four tables corresponding to odd sin, even sin, odd cos, and even cos. For example, using odd sin we obtain Table 3.6.1.4.

Table 3.6.1.4.

$\theta = 2\pi j/n$	$\sin\theta$	$\sin 3\theta$	$\sin 5\theta$
0	$A \times 0 \cdot 000 =$	$x \quad 0 \cdot 000 =$	$x \quad 0 \cdot 000 =$
30	$B \times 0 \cdot 500 =$	$x \quad 1 \cdot 000 =$	$x \quad 0 \cdot 500 =$
60	$C \times 0 \cdot 866 =$	$x \quad 0 \cdot 000 =$	$x - 0 \cdot 866 =$
90	$D \times 1 \cdot 000 =$	$x - 1 \cdot 000 =$	$x \quad 1 \cdot 000 =$

Step 5: multiply each value in Table 3.6.1.4 by the corresponding value in column 1 of Table 3.6.1.3 and put answer in space in Table 3.6.1.4. Sum and divide by 6. This gives the odd coefficients. Do the same for even sin, odd cos, and even cos.

3.6.2 Simplified method for twelve terms

Copy the data into the positions $x[0]$, $x[1]$, ..., $x[11]$ and follow the instructions of the arithmetic and recopying tables below.

	$x[0]$ ___	$x[1]$ ___	$x[2]$ ___	$x[3]$ ___	$x[4]$ ___	$x[5]$ ___	$x[6]$ ___
	$x[11]$ ___	$x[10]$ ___	$x[9]$ ___	$x[8]$ ___	$x[7]$ ___		
Add	$P[0]$ ___	$P[1]$ ___	$P[2]$ ___	$P[3]$ ___	$P[4]$ ___	$P[5]$ ___	$P[6]$ ___
Subtract		$Q[1]$ ___	$Q[2]$ ___	$Q[3]$ ___	$Q[4]$ ___	$Q[5]$ ___	

	$P[0]$ ___	$P[1]$ ___	$P[2]$ ___	$P[3]$ ___	$Q[1]$ ___	$Q[2]$ ___	$Q[3]$ ___
	$P[6]$ ___	$P[5]$ ___	$P[4]$ ___		$Q[3]$ ___	$Q[4]$ ___	
Add	$R[0]$ ___	$R[1]$ ___	$R[2]$ ___	$R[3]$ ___	$R[4]$ ___	$R[5]$ ___	$R[6]$ ___
Subtract	$S[0]$ ___	$S[1]$ ___	$S[2]$ ___		$S[4]$ ___	$S[5]$ ___	

	$R[0]$ ___	$R[1]$ ___	$R[4]$ ___	$S[4]$ ___
	$R[3]$ ___	$R[2]$ ___	$R[6]$ ___	$S[5]$ ___
Add	$T[0]$ ___	$T[1]$ ___	$T[2]$ ___	$T[3]$ ___
Subtract	$U[0]$ ___	$U[1]$ ___	$U[2]$ ___	$U[3]$ ___

Divide	2)$R[4]$ ___	2)$S[2]$ ___	2)$T[1]$ ___	2)$U[1]$ ___
	$V[0]$ ___	$V[1]$ ___	$V[2]$ ___	$V[3]$ ___

	$\log R[5]$ _____	$\log S[1]$ _____	$\log T[3]$ _____	$\log U[3]$ _____
$\mathrm{Log}\,0\cdot866$	$\bar{1}\cdot93753$	$\bar{1}\cdot93753$	$\bar{1}\cdot93753$	$\bar{1}\cdot93753$
Add	_____	_____	_____	_____
Antilogs	$W[0]$ _____	$W[1]$ _____	$W[2]$ _____	$W[3]$ _____

	$S[0]$ ___	$V[0]$ ___
	$V[1]$ ___	$R[6]$ ___
Add	$Y[0]$ ___	$Y[1]$ ___

	$T[0]$ ___	$U[0]$ ___		if $\bar{a}[6]$ positive	$\Phi[6]=0$
	$+T[1]$ ___	$-U[1]$ ___		negative	$\Phi[6]=\pi$
	12) ___	12) ___	$\bar{a}^2[6]=A^2[6]$ ___		
	$\bar{a}[0]$ ___	$\bar{a}[6]$ ___	$A[6]$ ___		$\Phi[6]$ ___
					$\Phi[6]/6$ ___

$Y[0]$ ___ $Y[1]$ ___ $a^2[1]$ ___ $\log b[1]$ ___
$+W[1]$ ___ $+W[0]$ ___ $+b^2[1]$ ___ $-\log a[1]$ ___
6)___ ___   6)___ ___ $A^2[1]$ ___ add   ___
$a[1]$ ___ $b[1]$ ___ $(A^2[1])^{\frac{1}{2}}$ ___ antilog arctan   ___

adjust to quadrant $\Phi[1]$ ___
$\Phi[1]/1$ ___

$U[0]$ ___ $a^2[2]$ ___ $\log b[2]$ ___
$+V[3]$ ___ $+b^2[2]$ ___ $-\log a[2]$ ___
6)___ ___   6) $W[2]$ ___ $A^2[2]$ ___ add   ___
$a[2]$ ___ $b[2]$ ___ $(A^2[2])^{\frac{1}{2}}$ ___ antilog arctan

adjust to quadrant $\Phi[2]$ ___
$\Phi[2]/2$ ___

$S[0]$ ___ $a^2[3]$ ___ $\log b[3]$ ___
$-S[2]$ ___ $+b^2[3]$ ___ $-\log a[3]$ ___
6)___ ___   6) $U[2]$ ___ $A^2[3]$ ___ add   ___
$a[3]$ ___ $b[3]$ ___ $(A^2[3])^{\frac{1}{2}}$ ___ antilog arctan   ___

adjust to quadrant $\Phi[3]$ ___
$\Phi[3]/3$ ___

$T[0]$ ___ $a^2[4]$ ___ $\log b[4]$ ___
$-V[2]$ ___ $+b^2[4]$ ___ $-\log a[4]$ ___
6)___ ___   6) $W[3]$ ___ $A^2[4]$ ___ add   ___
$a[4]$ ___ $b[4]$ ___ $(A^2[4])^{\frac{1}{2}}$ ___ antilog arctan   ___

adjust to quadrant $\Phi[4]$ ___
$\Phi[4]/4$ ___

$Y[0]$ ___ $Y[1]$ ___ $a^2[5]$ ___ $\log b[5]$ ___
$-W[1]$ ___ $-W[0]$ ___ $+b^2[5]$ ___ $-\log a[5]$ ___
6)___ ___   6)___ ___ $A^2[5]$ ___ add   ___
$a[5]$ ___ $b[5]$ ___ $(A^2[5])^{\frac{1}{2}}$ ___ antilog arctan   ___

adjust to quadrant $\Phi[5]$ ___
$\Phi[5]/5$ ___

$$\%\sigma^2[k] = 100\,\sigma^2[k]/\sigma^2[\text{total}]$$

$\sigma^2[1] = A^2[1]/2$ _____ $\%\sigma^2[1]$ _____

$\sigma^2[2] = A^2[2]/2$ _____ $\%\sigma^2[2]$ _____

$\sigma^2[3] = A^2[3]/2$ _____ $\%\sigma^2[3]$ _____

$\sigma^2[4] = A^2[4]/2$ _____ $\%\sigma^2[4]$ _____

$\sigma^2[5] = A^2[5]/2$ _____ $\%\sigma^2[5]$ _____

$\sigma^2[6] = A^2[6]$ _____ $\%\sigma^2[6]$ _____

$\sigma^2[\text{total}] = \text{sum}$ _____

Use Equation (3.2.7) to convert the phase shift to units of t.

3.7 An example of the use of phase

Very few variables are truly periodic; so plain harmonic analysis has limited use in isolating the unknown scale components in data. However, the technique may be considered from the point of view of curve fitting and, where an apparent cyclical process is acting, may be used to specify objectively the oscillation in terms of amplitude and phase. Of course, as in most curve-fitting problems it is not at all clear that the particular curve selected, in this case a single sinusoid, is necessarily the best. On the other hand, a sinusoid does describe a truly periodic process and computationally it is easily fitted.

A set of data which, it may be argued, are periodic is monthly normal precipitation (mean of the monthly totals for a standard 30 year period) because the normal for each month will be the same for every year. Furthermore, there are supporting physical arguments for the case of an annual and possibly of a semi-annual oscillation (Schwerdtfeger and Prohaska, 1956; Van Loon and Jeune, 1969) in precipitation. Consequently the fitting of a cosine curve with an annual period seems justified. The two resulting variables, amplitude and phase, turn out to be two very useful objective descriptions which may be mapped to show the annual variation of monthly precipitation. This application was first introduced by Horn and Bryson (1960) for the United States and has since been applied to Canada (Sabbagh and Bryson, 1962), Australia (Fitzpatrick, 1964), New Zealand (Rayner, 1965), and the British Isles (Rayner, 1966, unpublished). Here, as an example, the phase map for New Zealand is reproduced (Figure 3.7.1). As long-term means and normals were available for only about 500 stations, which were located mainly in the more densely populated areas, another 1200 stations with approximately 10 years of record each were added to the analysis. Many of the latter were considered to be unreliable, yet in most cases consistent results in the phase calculations permitted the draughting of isopleths accurately to within 10 miles except in the Southern Alps (dashed line) where data were still sparse. The map, as a description of the date of maximum precipitation as predicted by a fitted curve, speaks for itself, although it poses more

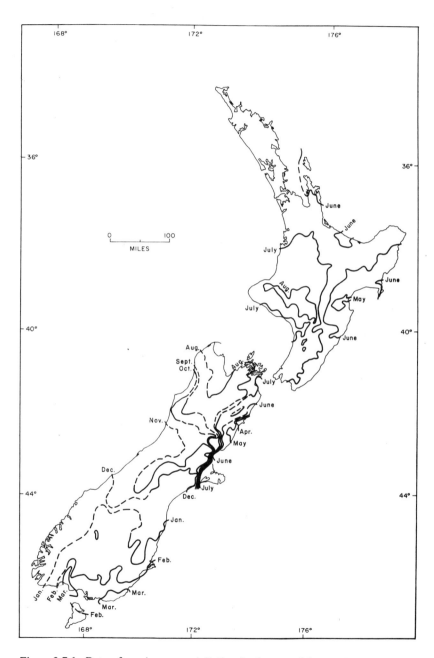

Figure 3.7.1 Date of maximum precipitation in the annual harmonic.

questions than it answers. For instance, it is clear that the date of the maximum passes from winter (June) in the north through spring to mid-summer (January) in the South. The obvious and simple explanation for this shift involves the expansion (in autumn) and contraction (in spring) of the southern hemisphere circumpolar atmospheric westerlies and the related movement of cyclone tracks. This is only a hypothesis, however, and it does not explain the west to east variation, particularly along the 44th parallel. It suggests the need for further research. Small amplitudes could give rise to error in phase estimation but they are greater than $\frac{1}{2}$ inch (one-inch range) in all areas except in a band running southwards from the central North Island along the eastern half of the Southern Alps. In general the amplitude map is similar to the total annual precipitation map. Maximum amplitudes of over 2 inches occur in all four corners of the North Island and in the Southern Alps.

Examination of the above references on cycles in monthly normal precipitation will reveal that harmonic analysis has been extended to other frequencies. Unfortunately it is not legitimate in this context to use the technique further on such data. It is a common practice to average all Januarys, Februarys, etc., but it forces harmonics of the annual period (i.e. frequencies of 2, 3, 4, 5, and 6 cycles per year) upon the results. The annual cycle itself is unaffected by the process because the least-squares fit to 30 years of monthly data is the same as the fit to one year of average monthly data. In fact, the preliminary process of calculating means is no different from calculating sums in the Buys–Ballot schedule of Schuster's periodogram (Conrad and Pollack, 1950, p.356). On the other hand, an 8 month cycle could occur in 30×12 months, and in fact there is no obvious reason why this should not be any more important than a 4 month cycle; yet the averaging process prevents it from being calculated. Furthermore, some of the variability from this frequency must appear in the calculable frequencies. Therefore, whereas harmonic coefficients may be obtained from such data and used in numerical manipulations, they may not be interpreted as having any physical meaning.

3.8 Exercises

Use the simplified method in Section 3.6.2 to obtain the basic results in the following from the data in Section 1.6, which are average January temperatures at approximately 5 feet above ground level in degrees fahrenheit. The sequence was obtained at $30°$ longitude intervals eastward along latitude $60°$ N with the origin on the 0 meridian.

3.8.1 Calculate the $a[k]$'s, $b[k]$'s, $A[k]$'s, and $\Phi[k]$'s. List the location (longitude) of the first maximum of each harmonic.

3.8.2 Plot the seven curves from the $A[k]$'s and $\Phi[k]$'s using $0·75$ inches for $\Delta\theta$ and a scale of $10°$F to 1 inch on the ordinate.

3.8.3 Calculate the $\hat{\sigma}^2[k]$'s and $\%\hat{\sigma}^2[k]$'s.

3.8.4 Plot the spectra of the a's, b's, and $\%\hat{\sigma}^2$'s.

3.8.5 Reading: Hamming (1962, pp.67–78); Horn and Bryson (1960); Peixoto *et al.* (1964); the section on hemispheric waves in Hare (1960), and look at the associated spectral diagrams in Wiin–Nielsen (1959) and Wiin–Nielsen *et al.* (1964).

4

Complex number representation

4.1 The exponential number

By definition the exponential number, which is represented by e, is the limit of the expression $(1+1/n)^n$ as n approaches infinity, i.e.

$$e = \lim_{n \to \infty} \left(1 + \frac{1}{n}\right)^n \tag{4.1.1}$$

$$= 2 \cdot 718 \dots . \tag{4.1.2}$$

This number has many uses particularly in calculus. For example,

$$\frac{d}{dt}(\log_{10} t) = \frac{1}{t} \log_{10} e \tag{4.1.3}$$

$$\int \frac{1}{t} dt = \log_e t + \text{constant} \tag{4.1.4}$$

$$\frac{d}{dt}(u^t) = u^t \log_e u . \tag{4.1.5}$$

Also

$$\frac{d}{dt}(e^{ut}) = u e^{ut} \tag{4.1.6}$$

$$\int e^{ut} dt = \frac{1}{u} e^{ut} + \text{constant} . \tag{4.1.7}$$

The abbreviated $\exp(uvt)$, which frequently replaces e^{uvt}, will be used in the following.

The functions $\exp(-t)$ and $\exp(t)$ are shown in Figure 4.1.1.

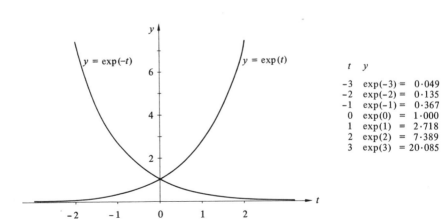

t	y	
-3	$\exp(-3) =$	$0 \cdot 049$
-2	$\exp(-2) =$	$0 \cdot 135$
-1	$\exp(-1) =$	$0 \cdot 367$
0	$\exp(0) =$	$1 \cdot 000$
1	$\exp(1) =$	$2 \cdot 718$
2	$\exp(2) =$	$7 \cdot 389$
3	$\exp(3) =$	$20 \cdot 085$

Figure 4.1.1.

4.2 Power series

4.2.1 Maclaurin's expansion

By Maclaurin's expansion a function $y[t]$ may be expanded as power series in t, i.e.

$$y[t] = c_0 + c_1 t + c_2 t^2 + c_3 t^3 + \ldots + c_n t^n + \ldots , \qquad (4.2.1.1)$$

where the coefficients are found by differentiation.

Differentiate Equation (4.2.1.1) to find

$$y'[t] = 1c_1 + 2c_2 t + 3c_3 t^2 + \ldots + nc_n t^{n-1} + \ldots$$

and again

$$y''[t] = \quad 2.1.c_2 + 3.2.c_3 t + \ldots + n.(n-1).c_n t^{n-2} + \ldots ,$$

yet again

$$y'''[t] \qquad\qquad 3.2.1c_3 + \ldots + n.(n-1).(n-2).c_n t^{n-3} + \ldots$$
$$\vdots$$
$$y^n[t] \qquad\qquad n!c_n + \ldots \qquad\qquad (4.2.1.2)$$

etc.

Put $t = 0$ then

$$y[0] = c_0$$
$$y'[0] = c_1 1!$$
$$y''[0] = c_2 2!$$
$$y'''[0] = c_3 3! \qquad\qquad (4.2.1.3)$$
$$\vdots$$
$$y^n[0] = c_n n! \qquad\text{etc.}$$

Put Equation (4.2.1.3) in Equation (4.2.1.1) and then

$$y[t] = y[0] + \frac{y'[0]t}{1!} + \frac{y''[0]t^2}{2!} + \ldots + \frac{y^n[0]t^n}{n!} . \qquad (4.2.1.4)$$

Now to convert a function into a power series its differentials must be obtained, t must be set equal to zero, and the results must be substituted into Equation (4.2.1.4).

4.2.2 The exponential as a power series

The differentials of e^t are

$$y[t] = e^t \qquad y[0] = 1$$
$$y'[t] = e^t \qquad y'[0] = 1$$
$$\vdots \qquad\qquad \vdots$$
$$y^n[t] = e^t \qquad y^n[0] = 1 .$$

Therefore

$$e^t = 1 + t + \frac{t^2}{2!} + \frac{t^3}{3!} + \frac{t^4}{4!} + ... + \frac{t^n}{n!} + ...$$ (4.2.2.1)

and similarly

$$e^{-t} = 1 - t + \frac{t^2}{2!} - \frac{t^3}{3!} + \frac{t^4}{4!} - ...$$ (4.2.2.2)

4.2.3 The sine function

$$y[t] = \sin t, \qquad y[0] = 0,$$
$$y'[t] = \cos t, \qquad y'[0] = 1,$$
$$y''[t] = -\sin t, \qquad y''[0] = 0,$$
$$y'''[t] = -\cos t, \qquad y'''[0] = -1.$$

Therefore

$$\sin t = t - \frac{t^3}{3!} + \frac{t^5}{5!} - \frac{t^7}{7!} + ...$$ (4.2.3.1)

The sine function is known as an odd function since it is asymmetrical around $t = 0$ but it can be seen here that it is odd also in the powers of t. Equation (4.2.3.1) may be used to compute the sine function where t is in radians.

4.2.4 The cosine function

$$y[t] = \cos t, \qquad y[0] = 1,$$
$$y'[t] = -\sin t, \qquad y'[0] = 0,$$
$$y''[t] = -\cos t, \qquad y''[0] = -1,$$
$$y'''[t] = \sin t, \qquad y'''[0] = 0,$$
$$y''''[t] = \cos t, \qquad y''''[0] = 1.$$

Therefore

$$\cos t = 1 - \frac{t^2}{2!} + \frac{t^4}{4!} - \frac{t^6}{6!} + ...$$ (4.2.4.1)

The cosine function is known as an even function.

4.3 The imaginary number and relationships between the exponential and sine and cosine

Let i be an imaginary number such that $i^2 = -1$ and then $\exp(it)$, from Equation (4.2.2.1), becomes

$$\exp(it) = 1 + it - \frac{t^2}{2!} - \frac{it^3}{3!} + \frac{t^4}{4!} + \frac{it^5}{5!} - \frac{t^6}{6!} - ... .$$ (4.3.1)

Rearrange Equation (4.3.1) to group separately the real and imaginary terms

$$\exp(it) = \left(1 - \frac{t^2}{2!} + \frac{t^4}{4!} - \frac{t^6}{6!} + \cdots\right) + i\left(t - \frac{t^3}{3!} + \frac{t^5}{5!} - \frac{t^7}{7!} \cdots\right). \tag{4.3.2}$$

The real terms given in the first bracket on the right of Equation (4.3.2) are equivalent to $\cos t$ [see Equation (4.2.4.1)] and the second bracket to $\sin t$ [see Equation (4.2.3.1)]. Therefore Equation (4.3.2) may be rewritten

$$\exp(it) = \cos t + i\sin t. \tag{4.3.3}$$

From Equation (4.2.2.2)

$$\exp(-it) = 1 - it - \frac{t^2}{2!} + \frac{it^3}{3!} + \frac{t^4}{4!} - \frac{it^5}{5!} - \frac{t^6}{6!} + \cdots. \tag{4.3.4}$$

Rearranging and substituting as in Equations (4.3.2) and (4.3.3) gives

$$\exp(-it) = \cos t - i\sin t. \tag{4.3.5}$$

Add Equation (4.3.5) to (4.3.3) and divide by 2, and then

$$\frac{\exp(it) + \exp(-it)}{2} = \cos t. \tag{4.3.6}$$

Subtract Equation (4.3.5) from (4.3.3) and divided by 2i, and then

$$\frac{\exp(it) - \exp(-it)}{2i} = \sin t. \tag{4.3.7}$$

Multiply top and bottom of the left-hand side of Equation (4.3.7) by i:

$$\frac{i\{\exp(-it) - \exp(it)\}}{2} = \sin t. \tag{4.3.8}$$

4.4 Fourier series in complex form

With the use of Equations (4.3.6) and (4.3.8), Equation (2.4.2) may now be rewritten

$$x[j] = \tfrac{1}{2}a[0] + \sum_{k=1}^{n/2} \left\{a[k]\cos\left(\frac{2\pi jk}{n}\right) + b[k]\sin\left(\frac{2\pi jk}{n}\right)\right\}$$

$$= \tfrac{1}{2}a[0] + \sum_{k=1}^{n/2} \left\{a[k]\frac{\exp(i2\pi jk/n) + \exp(-2\pi jk/n)}{2}\right.$$

$$\left. + b[k]\frac{i\{\exp(-i2\pi jk/n) - \exp(i2\pi jk/n)\}}{2}\right\} \tag{4.4.1}$$

and rearranging gives

$$x[j] = \tfrac{1}{2}a[0] + \sum_{k=1}^{n/2} \left\{\frac{a[k] - ib[k]}{2}\exp\left(\frac{i2\pi jk}{n}\right) + \frac{a[k] + ib[k]}{2}\exp\left(\frac{-i2\pi jk}{n}\right)\right\}$$

$$\tag{4.4.2}$$

Let

$$c[k] = \frac{a[k] - ib[k]}{2}, \qquad c[-k] = \frac{a[k] + ib[k]}{2}, \qquad \text{and} \quad c[0] = \frac{a[0] + i0}{2}$$

whence

$$x[j] = c[0] + \sum_{k=1}^{n/2} \left\{ c[k] \exp\left(\frac{i2\pi jk}{n}\right) + c[-k] \exp\left(\frac{-i2\pi jk}{n}\right) \right\}, \qquad (4.4.3)$$

which is the same as

$$x[j] = \sum_{k=-n/2}^{n/2} c[k] \exp\left(\frac{i2\pi jk}{n}\right) \qquad (4.4.4)$$

$$= \sum_{k=-n/2}^{n/2} \left(\frac{a[k] - ib[k]}{2}\right) \left\{ \cos\left(\frac{2\pi jk}{n}\right) + i\sin\left(\frac{2\pi jk}{n}\right) \right\} \qquad (4.4.5)$$

$$= \sum_{k=-n/2}^{n/2} \left[\left\{ \frac{a[k]}{2} \cos\left(\frac{2\pi jk}{n}\right) + \frac{b[k]}{2} \sin\left(\frac{2\pi jk}{n}\right) \right\} \right.$$
$$\left. + i\left\{ \frac{a[k]}{2} \sin\left(\frac{2\pi jk}{n}\right) - \frac{b[k]}{2} \cos\left(\frac{2\pi jk}{2}\right) \right\} \right], \qquad (4.4.6)$$

which are equivalent complex forms of Equation (2.4.2). If $x[j]$ is complex,

$$x[j] = x_R[j] + ix_I[j] ; \qquad (4.4.7)$$

real and imaginary parts of Equations (4.4.6) and (4.4.7) may be equated separately. If $x[j]$ is real, as is the usual case, Equation (4.4.6) reduces to Equation (2.4.2).

In the above, summation is conducted over positive and negative with the limits for even n. If n were odd, the limits would be $-(n-1)/2$ and $(n-1)/2$.

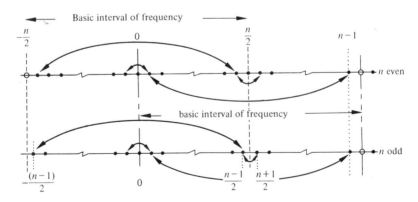

Figure 4.4.1.

As already mentioned in Chapter 3 the a's and b's are periodic, so summation could proceed from $k = 0$ to $k = n - 1$. Then

$$\frac{a[-1]}{2} = \frac{a[n-1]}{2} = \frac{a[1]}{2} \quad \text{and} \quad \frac{b[-1]}{2} = \frac{b[n-1]}{2} = -\frac{b[1]}{2} . \tag{4.4.8}$$

At k_{max}, $a[n/2]/2$ stands alone if n is even; otherwise $a[(n-1)/2]$ and $b[(n-1)/2]$ are matched at $[(n+1)/2]$ and $[-(n-1)/2]$ (see Figure 4.4.1).

Figure 4.4.1 also demonstrates why the formulae for the a's with subscripts $k = 0$ and $k = n/2$ in the one-sided spectrum differ from the other by a factor of two: they occur only once each in one complete basic interval of frequency, whereas the others occur twice.

4.5 Calculation of complex Fourier coefficients

Since $a[k]$ and $b[k]$ are still orthogonal, an equation equivalent to Equations (3.3.2)–(3.3.6) may be written for $\exp(\mathrm{i}2\pi kt/T)$:

$$\int_{-T/2}^{T/2} \exp\left(\frac{\mathrm{i}2\pi kt}{T}\right)\exp\left(-\frac{\mathrm{i}2\pi k't}{T}\right)\mathrm{d}t \quad \begin{aligned} &= 0 \text{ for } k \neq k' \\ &= T \text{ when } k = k' . \end{aligned} \tag{4.5.1}$$

To obtain a coefficient $c[k']$ multiply each side of the integral form of Equation (4.4.4) by $\exp(-\mathrm{i}2\pi k't/T)$ and integrate from $-T/2$ to $T/2$. Therefore

$$\int_{-T/2}^{T/2} x[t]\exp\left(-\frac{\mathrm{i}2\pi k't}{T}\right)\mathrm{d}t$$

$$= \int_{-T/2}^{T/2}\left\{\sum_{k=-n/2}^{n/2} c[k]\exp\left(\frac{\mathrm{i}2\pi kt}{T}\right)\exp\left(-\frac{\mathrm{i}2\pi k't}{T}\right)\right\}\mathrm{d}t \tag{4.5.2}$$

$$= \int_{-T/2}^{T/2} c[-n/2]\exp\left(-\frac{\mathrm{i}2\pi(n/2)t}{n\Delta t}\right)\exp\left(-\frac{\mathrm{i}2\pi k't}{n\Delta t}\right)\mathrm{d}t + \dots$$

$$+ \int_{-T/2}^{T/2} c[k']\exp\left(\frac{\mathrm{i}2\pi k't}{n\Delta t}\right)\exp\left(-\frac{\mathrm{i}2\pi k't}{n\Delta t}\right)\mathrm{d}t + \dots$$

$$+ \int_{-T/2}^{T/2} c[n/2]\exp\left(\frac{\mathrm{i}2\pi(n/2)t}{n\Delta t}\right)\exp\left(-\frac{\mathrm{i}2\pi k't}{n\Delta t}\right)\mathrm{d}t \tag{4.5.3}$$

$$= c[k']T . \tag{4.5.4}$$

Rearranging Equation (4.5.4) and dropping the prime gives

$$c[k] = \frac{1}{T}\int_{-T/2}^{T/2} x[t]\exp\left(-\frac{\mathrm{i}2\pi kt}{T}\right)\mathrm{d}t \tag{4.5.5}$$

or, in summation terms,

$$c[k] = \frac{1}{n}\sum_{j=-n/2}^{n/2} x[j]\exp\left(-\frac{\mathrm{i}2\pi jk}{n}\right) . \tag{4.5.6}$$

4.6 Geometrical representation

Whereas real numbers may be plotted along a straight line from an arbitrary origin (one dimension), complex numbers require a plane for their representation. The plane is known as the complex plane and the resulting diagram an Argand diagram, after Jean Robert Argand (see Figure 4.6.1).

In the Cartesian coordinate system of the complex plane the abscissa is known as the real axis and the ordinate the imaginary axis. Then the complex number $\alpha + i\beta$ is the point (α, β) and $\alpha - i\beta$ is point $(\alpha, -\beta)$.

The distance (vector) which joins the origin to (α, β) is known as the *modulus* and is defined as

$$|\alpha + i\beta| = +(\alpha^2 + \beta^2)^{\frac{1}{2}} , \tag{4.6.1}$$

which is non-zero and is a real number.

The *conjugate* of a complex number is obtained by changing the sign of i. Thus the conjugate of $\alpha - i\beta$ is $\alpha + i\beta$. Also note that

$$(\alpha - i\beta)(\alpha + i\beta) = \alpha^2 + \beta^2 . \tag{4.6.2}$$

Where only one letter, for example, c, is used to represent a complex number, its conjugate is denoted by an asterisk, c^*.

The complex number (α, β) may also be defined in terms of the angle θ, i.e.

$$\alpha = +(\alpha^2 + \beta^2)^{\frac{1}{2}} \cos\theta , \tag{4.6.3}$$

$$\beta = +(\alpha^2 + \beta^2)^{\frac{1}{2}} \sin\theta , \tag{4.6.4}$$

$$\alpha + i\beta = (\alpha^2 + \beta^2)^{\frac{1}{2}} (\cos\theta + i\sin\theta) \tag{4.6.5}$$

$$= (\alpha^2 + \beta^2)^{\frac{1}{2}} e^{i\theta} \tag{4.6.6}$$

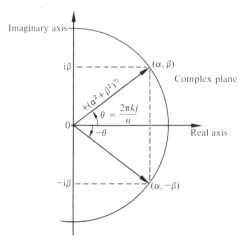

Figure 4.6.1. The Argand diagram.

Frequently $\cos\theta + i\sin\theta$ is written $\text{cis}\theta$. The corresponding complex spectra are therefore likewise known as cis spectra. These will be discussed in Section 4.8.

4.7 Graphical method of finding amplitudes

From Figure 4.6.1 and Equation (4.6.6) it is clear that the complex number function $(\alpha^2 + \beta^2)^{\frac{1}{2}}e^{i\theta}$ may be represented by a rotating vector of magnitude $(\alpha^2 + \beta^2)^{\frac{1}{2}}$.

If $x[t]$ is thought of as the motion of a point produced by the sum of such rotating vectors, a useful graphical solution for the amplitudes is available. To obtain a mathematical solution for a particular frequency the function was multiplied by the corresponding negative frequency $e^{-ik'2\pi t/T}$ and averaged [Equation (4.5.2)]. Here a rotation in the negative sense is applied, followed by averaging (see Figure 4.7.1). For example, vector A rotates at a rate of 1 cycle per basic interval T and vector B at 4 cycles per basic interval.

To find the amplitude of B the whole system is rotated at a rate equal and opposite to that of B, i.e. $-4/T$, a *negative* frequency.

The average of the path of P from the origin is equal to the amplitude of B.

Thus, in Figure 4.7.2 the example with added rotation $-4/T$, A rotates at $-3/T$ and B at $+4/T$. The average of P path in this system is the center of the circle.

In summary, then, the amplitude of B was found by rotating the whole system at an equal and opposite frequency to B and averaging.

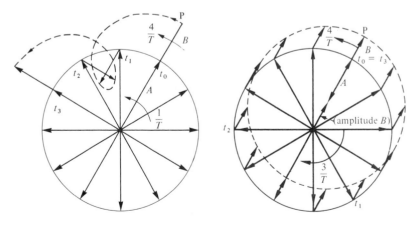

Figure 4.7.1.   **Figure 4.7.2.**

4.8 The complex or cis spectrum

4.8.1 Odd and even functions

Some mention has already been made of odd and even functions in Section 4.2 in connection with sines and cosines. Other descriptive words such as antisymmetrical and symmetrical might have been used to describe the arrangement of the functions around the origin. To make some of the following discussion simple it is now necessary to show that any function $x[t]$ may be separated into even or symmetrical, $x_E[t]$, and odd or antisymmetrical, $x_O[t]$, parts. Then, since even functions contain only cosines, Fourier analysis is simplified.

For the even function let

$$x_E[t] = \tfrac{1}{2}(x[t]+x[-t])$$ (4.8.1.1)

and for the odd function let

$$x_O[t] = \tfrac{1}{2}(x[t]-x[-t]) .$$ (4.8.1.2)

Then

$$x[t] = x_E[t]+x_O[t]$$ (4.8.1.3)

and

$$x_E[t] = a[0]+a[1]\cos\left(1\frac{2\pi t}{T}\right)+a[2]\cos\left(2\frac{2\pi t}{T}\right)+...+a[k]\cos\left(k\frac{2\pi t}{T}\right)+...$$

$$x_O[t] = \qquad b[1]\sin\left(1\frac{2\pi t}{T}\right)+b[2]\sin\left(2\frac{2\pi t}{T}\right)+...+b[k]\sin\left(k\frac{2\pi t}{T}\right)+... .$$
(4.8.1.4)

4.8.2 An example of division into odd and even functions

Let $x[t]$ be defined between $t = -6$ and $t = 6$ $(-6 \leqslant t \leqslant 6)$ (see Figure 4.8.2.1):

t	$=$	-6	-5	-4	-3	-2	-1	0	1	2	3	4	5	6
$x[t]$	$=$	6	7	1	2	0	8	4	2	6	10	7	5	4
$\pm t$	$=$	0	1	2	3	4	5	6						
$x[+t]$	$=$	4	2	6	10	7	5	4						
$x[-t]$	$=$	4	8	0	2	1	7	6						
$\tfrac{1}{2}(x[t]+x[-t]) =$		4	5	3	6	4	6	5						
$\tfrac{1}{2}(x[t]-x[-t]) =$			-3	3	4	3	-1	-1						

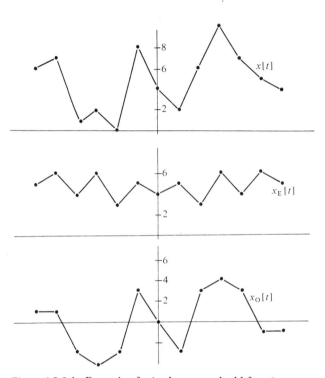

Figure 4.8.2.1. Example of mixed, even, and odd functions.

4.8.3 The combined sine and cosine spectra

In previous discussions (Section 4.5) the concept of integrating from
negative to positive time or space has been introduced without comment,
since it is not difficult to imagine negative time or space. However,
negative frequencies, introduced in Section 4.4, are not so easy to
interpret until they are associated with positive frequencies. The
usefulness of negative frequencies is apparent when an attempt is made to
plot the complete spectrum. The amplitudes $A[k]$ or variances $A^2[k]/2$
are even functions. Therefore they are symmetrical about zero frequency.
But they give only half the story and no information is given about the
phase. The alternative might be to plot separately $a[k]$ and $b[k]$. Then
$a[k]/2$ will be symmetrical with positive a's representing phases of zero
and negative a's representing phases of π. That the cosine function is
symmetrical in terms of a is seen if results of analysing data from 0 to
$+T/2$ are compared with those from 0 to $-T/2$. Similarly $b[k]/2$ is
antisymmetrical. A sine curve analysed from 0 to $-T/2$ will produce the
same amplitude with reversed phase (sign). This arrangement is also
apparent from Equations (4.3.6) and (4.3.8) for cosine and sine but it
will be noted that the sine amplitudes for positive frequencies are negative,
i.e. $b[k]/2$ will be plotted $-b[k]/2$ in a cis spectrum (Figure 4.8.3.1).

Amplitudes given in this form require two diagrams. It would be advantageous to have them both in one diagram. This may be accomplished, as shown in Figure 4.8.3.2, through the use of Equation (4.4.2) written

$$x_k[j] = \frac{a[k] + ib[k]}{2} \exp\left(-\frac{i2\pi kj}{n}\right) + \frac{a[k] - ib[k]}{2} \exp\left(\frac{i2\pi kj}{n}\right) . \quad (4.8.3.1)$$

The amplitude in the cis spectrum at $-k$ is therefore $(a[k] + b[k])/2$ and at $+k$, $(a[k] - b[k])/2$. a's and b's are obtainable from the cis spectrum using Equations (4.8.1.1) and (4.8.1.2).

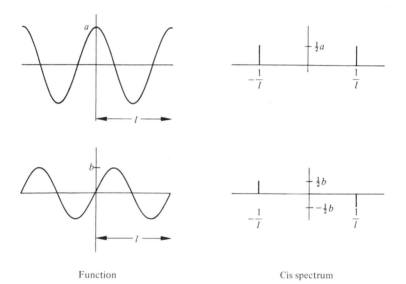

Function   Cis spectrum

Figure 4.8.3.1. Cosine and sine and their cis line spectra.

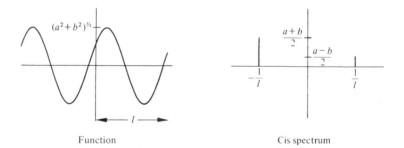

Function   Cis spectrum

Figure 4.8.3.2. Combined sine and cosine function and its combined cis line spectrum.

4.9 The fast Fourier transform algorithm

4.9.1 Basic algebra

This technique, which transforms a function into its Fourier components, was first introduced in a new form by Cooley and Tukey (1965) and given in detail by Gentleman and Sande (1966), although similar techniques had been used previously (Cooley $et\ al.$, 1967a). It reduces the number of calculations from a number proportional to n^2 to a number approximately proportional to $n \log n$ and has revolutionised spectral analysis, as will be seen later.

The basic equation from which the algorithm is developed is Equation (4.5.6) with summation over positive frequencies only:

$$nc[k] = \sum_{j=0}^{n-1} x[j] \exp\left(-\frac{i2\pi jk}{n}\right) . \tag{4.9.1.1}$$

Factors are then sought for n. For example, let

$$n = r_0 r_1 r_2$$

and

$$j = j_2 + r_2 j_1 + r_1 r_2 j_0 , \qquad k = k_0 + r_0 k_1 + r_0 r_1 k_2 ,$$

where j's and k's are integers and

$$0 \leqslant j_0, k_0 \leqslant r_0 - 1$$
$$0 \leqslant j_1, k_1 \leqslant r_1 - 1$$
$$0 \leqslant j_2, k_2 \leqslant r_2 - 1 .$$

Then Equation (4.9.1.1) becomes

$$nc[k_0 + r_0 k_1 + r_0 r_1 k_2] = \sum_{j_2=0}^{r_2-1} \sum_{j_1=0}^{r_1-1} \sum_{j_0=0}^{r_0-1} x[j_2 + r_2 j_1 + r_1 r_2 j_0]$$
$$\times \exp\left[-\frac{i2\pi}{n}(\{k_0 + r_0 k_1 + r_0 r_1 k_2\}\{j_2 + r_2 j_1 + r_1 r_2 j_0\})\right] . \tag{4.9.1.2}$$

If $\exp[-2\pi i(...)]$ is written $\mathcal{E}(...)$ then the exponential term expands to

$$\mathcal{E}\left(\frac{k_2(j_2 + r_2 j_1 + r_1 r_2 j_0)}{r_2} + \frac{k_1(j_2 + r_2 j_1 + r_1 r_2 j_0)}{r_1 r_2} + \frac{k_0(j_2 + r_2 j_1 + r_1 r_2 j_0)}{r_0 r_1 r_2}\right)$$

$$= \mathcal{E}\left(\frac{k_2 j_2}{r_2} + j_1 + r_1 j_0 + \left\{\frac{k_1(j_2 + r_2 j_1)}{r_1 r_2}\right\} + j_0 + \left\{\frac{k_0(j_2 + r_2 j_1 + r_1 r_2 j_0)}{r_0 r_1 r_2}\right\}\right)$$

but, since $\mathcal{E}(\text{integer}) = \exp(-2\pi i \times \text{integer}) = 1$ and all j's, k's, and r's are integers, three of the terms vanish and the last line becomes

$$\mathcal{E}\left(\frac{k_2 j_2}{r_2}\right) \mathcal{E}\left(\frac{k_1(j_2 + r_2 j_1)}{r_1 r_2}\right) \mathcal{E}\left(\frac{k_0(j_2 + r_2 j_1 + r_1 r_2 j_0)}{r_0 r_1 r_2}\right) .$$

Now terms which do not include the summation variable may be taken outside the summation sign. Therefore Equation (4.9.1.2) may be rewritten

$$nc[k_0+r_0k_1+r_0r_1k_2] = \sum_{j_2=0}^{r_2-1}\left[\sum_{j_1=0}^{r_1-1}\left\{\sum_{j_0=0}^{r_0-1}x[j_2+r_2j_1+r_1r_2j_0]\right.\right.$$

$$\left.\left.\times \text{\&}\left(\frac{k_0(j_2+r_2j_1+r_1r_2j_0)}{r_0r_1r_2}\right)\right\}\text{\&}\left(\frac{k_1(j_2+r_2j_1)}{r_1r_2}\right)\right]\text{\&}\left(\frac{k_2j_2}{r_2}\right).$$

$$(4.9.1.3)$$

This is a nested series of Fourier analyses or discrete Fourier transforms.

Step 1: consider the transform in brackets $\{...\}$ in Equation (4.9.1.3). This involves the summation of r_0 terms but it must be recalculated for each of the included variables other than j_0. That is, a separate sum will be produced for each value of $k_0(0, 1, ..., r_0-1)$, and $(j_2+r_2j_1)$ which varies from 0 to r_1r_2-1. In other words, step one can be thought of as producing r_0 rows each labelled by k_0. Within each row are r_1r_2 items each labelled with $(j_2+r_2j_1)$ as shown in Figure 4.9.1.1.

Each item in the matrix is the result of one transformation (multiplication and summation). The resulting new terms x' have the label k_0, $j_2+r_2j_1$, or $j_2+r_2j_1+r_1r_2k_0$ as can be seen from above. Therefore the equation at the end of this step is

$$nc[k_0+r_0k_1+r_0r_1k_2] = \sum_{j_2=0}^{r_2-1}\left[\sum_{j_1=0}^{r_1-1}x'[j_2+r_2j_1+r_1r_2k_0]\right.$$

$$\left.\times \text{\&}\left(\frac{k_1(j_2+r_2j_1)}{r_1r_2}\right)\right]\text{\&}\left(\frac{k_2j_2}{r_2}\right). \qquad (4.9.1.4)$$

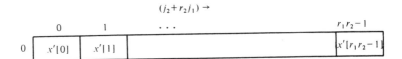

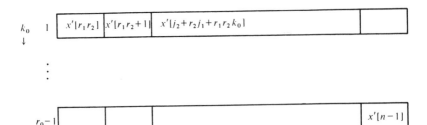

Figure 4.9.1.1. Matrix resulting from first transform.

Step 2: consider the transform in brackets $[\![\ldots]\!]$ in Equation (4.9.1.4). This involves the summation of r_1 terms but it must be recalculated for each of the included variables other than j_1. That is, a separate sum will be produced for each value of $k_1(0, 1, \ldots, r_1 - 1)$ and $j_2(0, 1, \ldots, r_2 - 1)$ for each of the $k_0(0, 1, \ldots, r_0 - 1)$ rows. In other words, step two can be thought of as producing r_1 new rows, each labelled by k_1, in each of the original r_0 rows. Within each new row are r_2 items each labelled with j_2. Each item is the result of another transformation. The resulting new terms x'' have the label $j_2 + r_2 k_1 + r_1 r_2 k_0$. Therefore the equation at the end of this step is

$$nc[k_0 + r_0 k_1 + r_0 r_1 k_2] = \sum_{j_2 = 0}^{r_2 - 1} x''[j_2 + r_2 k_1 + r_1 r_2 k_0] \& \left(\frac{k_2 j_2}{r_2} \right) . \quad (4.9.1.5)$$

Step 3: the last transform produces $r_0 r_1$ column matrices of r_2 rows labelled k_2. This yields the answer $c[k_2 + r_2 k_1 + r_1 r_2 k_0]$ which has reversed subscripts. A final step involves sorting the c's into serial order.

It should now be obvious that this process is not limited to three factors. For each one, $r_0, r_1, \ldots, r_m$, the data, originally in one row, are subdivided into more and more rows until only a column remains.

Step 1: r_0 rows.

Step 2: r_1 rows from each of the r_0 rows $= r_0 r_1$ rows.

Step 3: r_2 rows from each of the $r_0 r_1$ rows $= r_0 r_1 r_2$ rows.

$$\vdots \qquad \vdots$$

Last: $r_0 r_1 r_2 \ldots r_m = n$ rows.

For example, see Figure 4.9.1.2.

The problem is simplified if the factors are all equal,

$$r_0 = r_1 = r_2 = \ldots = r_m .$$

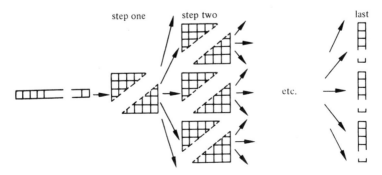

Figure 4.9.1.2.

Then each partial transform requires the same number of terms and, so long as counters can keep track of the labels and the argument of the exponential, a simple computer loop will suffice for the whole transform. The factor 4 turns out to be the most economical one although others may be used and have been included in a program written by Sande (1966 personal communication).

4.9.2 The computer loops

In constructing his computer program Sande allowed for transforming two series at once, $x[j]$ and $y[j]$, by assuming that $x[j]$ is real and $y[j]$ is imaginary. Then Equation (4.9.1.1) becomes

$$\alpha[k] + i\beta[k] = \sum_{j=0}^{n-1} (x[j] + iy[j]) \exp\left(-\frac{i2\pi jk}{n}\right) . \tag{4.9.2.1}$$

The transform in step one above is then

$$\sum_{j_0=0}^{r_0-1} (x[j_2 + r_2 j_1 + r_1 r_2 j_0] + iy[j_2 + r_2 j_1 + r_1 r_2 j_0])$$

$$\times \exp\left[-2\pi i \left\{\frac{k_0(j_2 + r_2 j_1 + r_1 r_2 j_0)}{r_0 r_1 r_2}\right\}\right] .$$

The exponential becomes

$$\cos\left\{\frac{2\pi k_0(j_2 + r_2 j_1 + r_1 r_2 j_0)}{r_0 r_1 r_2}\right\} - i\sin\left\{\frac{2\pi k_0(j_2 + r_1 j_1 + r_1 r_2 j_0)}{r_0 r_1 r_2}\right\} .$$

Then multiplying and equating real and imaginary parts gives

$$x'[j_2 + r_2 j_1 + r_1 r_2 k_0]$$

$$= \sum_{j_0=0}^{r_0-1} \left[x[j_2 + r_2 j_1 + r_1 r_2 j_0] \cos\left\{\frac{2\pi k_0(j_2 + r_2 j_1 + r_1 r_2 j_0)}{r_0 r_1 r_2}\right\}\right.$$

$$\left. + y[j_2 + r_2 j_1 + r_1 r_2 j_0] \sin\left\{\frac{2\pi k_0(j_2 + r_2 j_1 + r_1 r_2 j_0)}{r_0 r_1 r_2}\right\}\right] \tag{4.9.2.2}$$

and

$$y'[j_2 + r_2 j_1 + r_1 r_2 k_0]$$

$$= \sum_{j_0=0}^{r_0-1} \left[y[j_2 + r_2 j_1 + r_1 r_2 j_0] \cos\left\{\frac{2\pi k_0(j_2 + r_2 j_1 + r_1 r_2 j_0)}{r_0 r_1 r_2}\right\}\right.$$

$$\left. - x[j_2 + r_2 j_1 + r_1 r_2 j_0] \sin\left\{\frac{2\pi k_0(j_2 + r_2 j_1 + r_1 r_2 j_0)}{r_0 r_1 r_2}\right\}\right] . \tag{4.9.2.3}$$

Now, for example, let $r_0 = 4$. This produces four rows each labelled by k_0. There will be $n/4$ items, or columns each labelled $j_2 + r_2 j_1$, in each row. Consider Equation (4.9.2.2) which gives the general terms in

$$j_0 \to$$

$j_2+r_2j_1$ $\downarrow$		0	1	2	3
	0	0	$0+n/4$	$0+2n/4$	$0+3n/4$
	1	1	$1+n/4$	$1+2n/4$	$1+3n/4$
	·	·			·
	·	·	$[j_2+r_2j_1+r_1r_2j_0]$		·
	·	·			·
	$n/4$	$n/4$	$n/4+n/4$	$n/4+2n/4$	$n/4+3n/4$

Figure 4.9.2.1.

$$j_0 \to$$

$j_2+r_2j_1$ $\downarrow$				0	1	2	3
	0	$k_0\downarrow$	0	0	0	0	0
			1	$0+0$	$0+\dfrac{2\pi}{4}$	$0+\dfrac{2\pi2}{4}$	$0+\dfrac{2\pi3}{4}$
			2	$0+0$	$0+\dfrac{2\pi2}{4}$	$0+\dfrac{2\pi4}{4}$	$0+\dfrac{2\pi6}{4}$
			3	$0+0$	$0+\dfrac{2\pi3}{4}$	$0+\dfrac{2\pi6}{4}$	$0+\dfrac{2\pi9}{4}$
	1	$k_0\downarrow$	0	0	0	0	0
			1	$\dfrac{2\pi}{n}+0$	$\dfrac{2\pi}{n}+\dfrac{2\pi}{4}$	$\dfrac{2\pi}{n}+\dfrac{2\pi2}{4}$	$\dfrac{2\pi}{n}+\dfrac{2\pi3}{4}$
			2	$\dfrac{2\pi2}{n}+0$	$\dfrac{2\pi2}{n}+\dfrac{2\pi2}{4}$	$\dfrac{2\pi2}{n}+\dfrac{2\pi4}{4}$	$\dfrac{2\pi2}{n}+\dfrac{2\pi6}{4}$
			3	$\dfrac{2\pi3}{n}+0$	$\dfrac{2\pi3}{n}+\dfrac{2\pi3}{4}$	$\dfrac{2\pi3}{n}+\dfrac{2\pi6}{4}$	$\dfrac{2\pi3}{n}+\dfrac{2\pi9}{4}$
	·	·	·	·	·	·	·
	$n/4$	$k_0\downarrow$	0	0	0	0	0
			1	$\dfrac{2\pi}{4}+0$	$\dfrac{2\pi}{4}+\dfrac{2\pi}{4}$	$\dfrac{2\pi}{4}+\dfrac{2\pi2}{4}$	$\dfrac{2\pi}{4}+\dfrac{2\pi3}{4}$
			2	$\dfrac{2\pi2}{4}+0$	$\dfrac{2\pi2}{4}+\dfrac{2\pi2}{4}$	$\dfrac{2\pi2}{4}+\dfrac{2\pi4}{4}$	$\dfrac{2\pi2}{4}+\dfrac{2\pi6}{4}$
			3	$\dfrac{2\pi3}{4}+0$	$\dfrac{2\pi3}{4}+\dfrac{2\pi3}{4}$	$\dfrac{2\pi3}{4}+\dfrac{2\pi6}{4}$	$\dfrac{2\pi3}{4}+\dfrac{2\pi9}{4}$

Figure 4.9.2.2.

the sum. From it subscripts for x and y may be put down in tabular form as a function of the variables j_0 and $j_2 + r_2 j_1$ (k_0 does not appear), e.g. see Figure 4.9.2.1.

Similarly the angle may be listed as a function of j_0, k_0, and $j_2 + r_2 j_1$

if $\dfrac{2\pi k_0(j_2 + r_2 j_1 + r_1 r_2 j_0)}{4 r_1 r_2}$ is written $k_0\left\{\dfrac{2\pi(j_2 + r_2 j_1)}{n}\right\} + k_0\left(\dfrac{2\pi j_0}{4}\right)$

(see Figure 4.9.2.2.)

In general the angle may be written $k_0\theta + k_0 j_0(\pi/2)$, where $\theta = 2\pi(j_2 + r_2 j_1)/n$. Then since

$$\cos\left(k_0\theta + k_0 j_0\frac{\pi}{2}\right) = \cos k_0\theta \cos\left(k_0 j_0\frac{\pi}{2}\right) - \sin k_0\theta \sin\left(k_0 j_0\frac{\pi}{2}\right)$$

and

$\frac{\pi}{2}k_0 j_0$	$\frac{\pi}{2}$	$\frac{\pi}{2}2$	$\frac{\pi}{2}3$	$\frac{\pi}{2}4$	$\frac{\pi}{2}5$	$\frac{\pi}{2}6$	$\frac{\pi}{2}7$	$\frac{\pi}{2}8$	$\frac{\pi}{2}9$
$\cos\left(k_0 j_0\frac{\pi}{2}\right)$	0	−1	0	1	0	−1	0	1	0
$\sin\left(k_0 j_0\frac{\pi}{2}\right)$	1	0	−1	0	1	0	−1	0	1

the cosine term may be written for any block as in Figure 4.9.2.3.

$j_0 =$	0	1	2	3
$k_0 = 0$	1	1	1	1
1	$\cos\theta$	$-\sin\theta$	$-\cos\theta$	$\sin\theta$
2	$\cos 2\theta$	$-\cos 2\theta$	$\cos 2\theta$	$-\cos 2\theta$
3	$\cos 3\theta$	$\sin 3\theta$	$-\cos 3\theta$	$-\sin 3\theta$

Figure 4.9.2.3.

Similarly for

$$\sin\left(k_0\theta + k_0 j_0\frac{\pi}{2}\right) = \sin k_0\theta \cos\left(k_0 j_0\frac{\pi}{2}\right) + \cos k_0\theta \sin\left(k_0 j_0\frac{\pi}{2}\right)$$

we have the values shown in Figure 4.9.2.4.

$j_0 =$	0	1	2	3
$k_0 = 0$	0	0	0	0
1	$\sin\theta$	$\cos\theta$	$-\sin\theta$	$-\cos\theta$
2	$\sin 2\theta$	$-\sin 2\theta$	$\sin 2\theta$	$-\sin 2\theta$
3	$\sin 3\theta$	$-\cos 3\theta$	$-\sin 3\theta$	$\cos 3\theta$

Figure 4.9.2.4.

This allows the construction of a simple computer loop to complete Equations (4.9.2.2) and (4.9.2.3). With each pass through the loop, θ is incremented by adding one to the number inside the brackets, i.e. $(j_2 + r_2 j_1)$.

Equations corresponding to the transform in step two written in the form of (4.9.2.2) and (4.9.2.3) are

$$x''[j_2 + r_2 k_1 + r_1 r_2 k_0]$$
$$= \sum_{j_1=0}^{r_1-1} \left\{ x'[j_2 + r_2 j_1 + r_1 r_2 k_0] \cos\left[\frac{2\pi k_1(j_2 + r_2 j_1)}{r_1 r_2}\right] \right.$$
$$\left. + y'[j_2 + r_2 j_1 + r_1 r_2 k_0] \sin\left[\frac{2\pi k_1(j_2 + r_2 j_1)}{r_1 r_2}\right] \right\} \qquad (4.9.2.4)$$

and

$$y''[j_2 + r_2 k_1 + r_1 r_2 k_0]$$
$$= \sum_{j_1=0}^{r_1-1} \left\{ y'[j_2 + r_2 j_1 + r_1 r_2 k_0] \cos\left[\frac{2\pi k_1(j_2 + r_2 j_1)}{r_1 r_2}\right] \right.$$
$$\left. - x'[j_2 + r_2 j_1 + r_1 r_2 k_0] \sin\left[\frac{2\pi k_1(j_2 + r_2 j_1)}{r_1 r_2}\right] \right\} . \qquad (4.9.2.5)$$

If r_1 also equals 4, the procedure is similar to step one. This produces four rows each labelled by k_1. There will be $n/(4 \times 4)$ items, or columns each labelled j_2, in each row, but also there will be $r_0 = 4$ sets, or rows, of these each labelled j_0.

Each set will produce a figure similar to Equation (4.9.2.1) with variables j_1 and j_2. The sets will be different by an increment of $n/4$. The angles (Figure 4.9.2.2) also will be similar with variables j_1, k_1, and j_2. The only difference will be that n is reduced to $n/4$ and j_2 is limited to $n(4 \times 4)$. Figures 4.9.2.3 and 4.9.2.4 will remain the same with $j_1 = j_0$, $k_1' = k_0$, and $\theta = 2\pi j_2/(n/4)$.

Consequently, the loop for the second transform (step two) is the same as the first with $n = n/4$ and allowance for more than one row, i.e. another loop. So long as 4 is the factor the loop may be used for successive transforms.

Thus Sande's loops are

$M = n$
10 $MR = M$ (start of loop for each step)
   test for 4 as factor
$M = M/4$
   $DO\ 30\ J = 1, M$ (start of loop for each column)
      $\theta = 2\pi(J-1)/MR$
      $DO\ 20\ K = MR, n, MR$ (start of loop for each row previously
                                                         produced)

$$J0 = J+K-MR,\ J1 = J0+M,\ J2 = J1+M,\ J3 = J2+M$$
$$R0 = X(J0)+X(J2)\quad R1 = X(J0)-X(J2)$$
$$I0 = Y(J0)+Y(J2)\quad I1 = Y(J0)-Y(J2)$$
$$R2 = X(J1)+X(J3)\quad R3 = X(J1)-X(J3)$$
$$I2 = Y(J1)+Y(J3)\quad I3 = Y(J1)-Y(J3)\quad \text{calculation of new}$$
$$X(J0) = R0+R2\qquad Y(J0) = I0+I2\qquad \text{items, one for each}$$
$$X(J2) = (R1+I3)\cos\theta + (I1-R3)\sin\theta\qquad \text{row 4 for } x$$
$$Y(J2) = (I1-R3)\cos\theta - (R1-I3)\sin\theta\qquad \text{4 for } y$$
$$X(J1) = (R0-R2)\cos 2\theta + (I0-I2)\sin 2\theta$$
$$Y(J1) = (I0-I2)\cos 2\theta - (R0-R2)\sin 2\theta$$
$$X(J3) = (R1-I3)\cos 3\theta + (I1+R3)\sin 3\theta$$
$$Y(J3) = (I1-R3)\cos 3\theta - (R1-I3)\sin 3\theta$$

20   continue
30   continue
   return to 10

Loops for other factors may be produced in the same way followed by final unscrambling.

It will be seen from Equations (4.9.2.4), etc., that the final series α and β are a mixture of a_x, b_x, a_y, and b_y. In fact, α from Equation (4.9.2.2) is a mixture of a_x and b_y; β from Equation (4.9.2.3) is a mixture of a_y and b_x. In each case the a's are even and the b's are odd and so Equations (4.8.1.1) and (4.8.1.2) will separate them.

The series α and β are calculated for $k = 0$ to $k = n - 1$; so the folding point is $n/2$ (see Figure 4.4.1).

Gentleman and Sande (1966) have pointed out that series too long for computer immediate access storage are easily accommodated by grouping the data into blocks. The first block should contain 1st, mth, $2m$th, ... , etc. data, and the second block the 2nd $(m+1)$th, $(2m+1)$th, ... , etc. data. Each is then transformed from a row to a column (Figure 4.9.1.2). Each column side by side forms a series of new rows each to be transformed to produce the final answer. The process of rearranging the data from columns in separate blocks to rows in separate blocks is the equivalent of a matrix transpose.

4.10 Exercises

4.10.1 Use Equations (3.3.11) and (3.3.12) to calculate $a[11]$ and $b[11]$ from the data in Exercise 1.6.1. Compare these with $a[1]$ and $b[1]$ and consider the symmetry of Figure 4.4.1.

4.10.2 List the cis spectrum $k = -n/2$ to $k = n/2$ of the results in Exercise 3.8.1. Note that this spectrum contains one more frequency than there are original points. Why is this? What is the alternative?

4.10.3 Write or obtain a computer program to calculate the one-dimensional Fourier transform from expirical data. Test the accuracy of the program by applying the transformation twice [i.e. use Equation (4.5.6) and then (4.4.4)].

4.10.4 Use the $a[k]$'s and $b[k]$'s from Exercise 3.8.1 to interpolate for every 2 degrees longitude. With the fast Fourier transform algorithm this may be accomplished by increasing the number of a's and b's from 12 to 180 through the insertion of zero magnitudes for the frequencies $k = 9$ to 183 before retransformation. (Half of $a[7]$ must appear at $k = 7$ and half at $k = 184$.) This increases the number of final estimates yet does not add further variance. Plot the results with those of Exercise 1.6.1.

4.10.5 Reading: Tolstov (1962, chapter 1); Gentleman and Sande (1966); Cooley *et al.* (1965); Singleton (1967).

5

The Fourier integral

5.1 Introduction

Besides the three restrictions listed in Section 3.1 the discussion so far has been limited to a particular class of function, the periodic function. The equations for the isolation of the frequency components in $x[t]$ have been given in terms of sine and cosine and in terms of complex numbers. Of course, the same analysis could be applied to non-periodic functions but the problem arises in interpreting the results.

Basically the function for analysis should be defined between $-\infty$ and $+\infty$. This requirement is seldom fulfilled and only a relatively short record (set of data) is available. Consequently, some theoretical history and future must be invented if the analysis is to proceed.

Two alternatives exist.

1 Assume true periodicity and apply harmonic analysis as already outlined. Integration for the $a[k]$ and $b[k]$ values is then necessary only in the basic interval 0 to 2π (or $-\pi$ to π, 0 to T or $-T/2$ to $T/2$).

2 Assume that the function is zero outside the interval for which data are available. Integration then proceeds from $-\infty$ to ∞ giving the so-called 'Fourier integral'. The results of analysis following such an assumption are derived and discussed in this section.

5.2 Stretching the basic interval

The solution for the case of zero outside the interval of information may be found through initially assuming that periodicity occurs after a certain interval $n\Delta t$ and then allowing n to approach infinity.

Assume that the function is even and is defined between $-\frac{1}{2}n\Delta t$ and $+\frac{1}{2}n\Delta t$ (see Figure 5.2.1).

Apply Equation (3.3.11) with limits $-n/2$ to $n/2$

$$a[k] = \frac{2}{n} \sum_{j=-n/2}^{n/2} x[j]\cos\left(\frac{2\pi kj}{n}\right)$$

(5.2.1)

or in integral form

$$a[k] = \frac{2}{T}\int_{-T/2}^{T/2} x[t]\cos\left(\frac{2\pi kt}{n\Delta t}\right)dt .$$

(5.2.2)

There are no $b[k]$ values. Assume that this analysis of Figure 5.2.1 produces a line spectrum as given in Figure 5.2.2.

It will be seen that the amplitude may take on negative values to allow for cosine waves which have phases of π radians. In other words the cosine curve is still symmetrical about zero but it starts at a negative ordinate value.

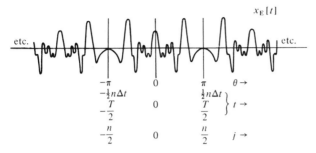

Figure 5.2.1.

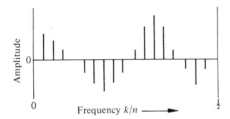

Figure 5.2.2.

Also, if frequency is measured in cycles per data spacing, k/n, the highest frequency calculable $(k/n)_{max}$ is $\frac{1}{2}$ (Section 3.2). This means that the length of the basic interval has no effect upon the limits of the spectrum (0 to $\frac{1}{2}$ cycles per data interval). Consequently, if the periodic interval in the process is doubled to $n'\Delta t = 2n\Delta t$ by inserting zeros (Figure 5.2.3), the spectrum will cover the same range.

On the other hand, the number of separate estimates of amplitudes will be doubled, since $k'_{max} = n'/2 = 2n/2 = 2k_{max}$.

What happens to the amplitudes?

$$a'[k] = \frac{2}{n'} \sum_{j=-n'/2}^{n'/2} x[j] \cos\left(\frac{2\pi k'j}{n'}\right) \tag{5.2.3}$$

$$= \frac{2}{2n} \sum_{j=-n}^{n} x[j] \cos\left(\frac{2\pi k'j}{2n}\right). \tag{5.2.4}$$

The estimate in this spectrum corresponding to k is $k' = 2k$. Therefore

$$a'[k] = \frac{2}{2n} \sum_{j=-n}^{n} x[j] \cos\left(\frac{2\pi kj}{n}\right). \tag{5.2.5}$$

Since $x[j] = 0$ for $j < -n/2$ and $j > n/2$, this may be written

$$a'[k] = \frac{2}{2n} \sum_{j=-n/2}^{n/2} x[j] \cos\left(\frac{2\pi kj}{n}\right) \tag{5.2.6}$$

which is half Equation (5.2.1). Therefore

$$a'[k] = \tfrac{1}{2}a[k] \ . \tag{5.2.7}$$

Therefore the amplitudes are halved and become twice as frequent as the basic interval is doubled.

If the basic interval is extended indefinitely, then the amplitude line, whilst being proportionately reduced, will merge to form a continuous curve (Figure 5.2.4).

Since the integral in Equation (5.2.2) remains unchanged as $n \to \infty$, the only problem remaining is that of the coefficient $2/n\Delta t$. This may be removed by absorbing it in $a[k]$, i.e. let

$$a[f] = \frac{n\Delta t a[k]}{2} = \int_{-\infty}^{\infty} x_E[t]\cos\left(\frac{2\pi kt}{n\Delta t}\right)\mathrm{d}t \ . \tag{5.2.8}$$

What is $a[f]$? $a[k]$ is an amplitude. $1/n\Delta t$ is a frequency increment. Therefore $n\Delta t a[k]$ is amplitude per frequency increment or amplitude density. So $a[f]$ is half the amplitude density at frequency f. Also, since $a[f]$ is a symmetrical function, the other half of the amplitude density is to be found at $a[-f]$. The relationship between amplitude density and amplitude is analogous to that between probability density and probability.

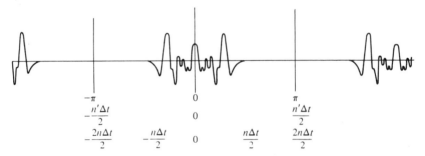

Figure 5.2.3.

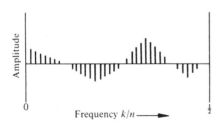

Figure 5.2.4.

The general equation for amplitude density now reads

$$a[f] = \int_{-\infty}^{\infty} x[t] \cos\left(\frac{2\pi kt}{n\Delta t}\right) dt \ .$$ (5.2.9)

Whilst n increases, k does also, so that k/n may be absorbed into one frequency term. There are two alternatives:

$$f = \frac{k}{n} \text{ cycles per data spacing}$$ (5.2.10)

$$f = \frac{k}{n\Delta t} \text{ cycles per units of } t \ .$$ (5.2.11)

If the second of these alternatives is taken, then the general equation for the amplitude density may be written

$$a[f] = \int_{-\infty}^{\infty} x[t] \cos(2\pi ft) dt$$ (5.2.12)

where $a[f]$ is known as the cosine transform of $x[t]$.

Usually in practice $x[t]$ will be a discrete variable $x[j]$ so that it may be calculated only for discrete frequencies k. Thus Equation (5.2.12) reverts to Equation (5.2.1). However, $a[k]$ no longer represents a discrete line but refers to a band $1/n\Delta t$ or $1/T$ wide. It may be depicted in the form of a histogram where the height of a band, instead of the length of the line in the discrete spectrum, is the magnitude of $a[k]$. As such it is a one-sided spectrum. Each $a[k]$ except $a[0]$ and $a[n/2]$ must be halved for the two-sided spectrum. If a density plot is required, the $a[k]$ values must be multiplied by $n\Delta t/2$ ($= T/2$) for the two-sided spectrum and the units of the ordinate will be in units of $x[j]$ times the units of t.

A similar argument may be applied to the odd function $x_0[t]$ to give values of $b[f]$, i.e.

$$b[f] = \int_{-\infty}^{\infty} x[t] \sin(2\pi ft) dt \ ,$$ (5.2.13)

where $b[f]$ is known as the sine transform of $x[t]$.

In Section 4.4 $c[k]$ was defined as a two-sided function equal to $(a[k] - ib[k])/2$. Under the same argument as was used for $a[f]$ and $b[f]$, $c[f]$ must be a two-sided amplitude density given by

$$c[f] = n\Delta t c[k] = a[f] - ib[f] \ .$$ (5.2.14)

The original function may now be written in the same form as Equation (2.4.2):

$$x[t] = \int_{-\infty}^{\infty} \{a[f] \cos(2\pi ft) + b[f] \sin(2\pi ft)\} df \ .$$ (5.2.15)

Furthermore, if capital letters, instead of c, are used to denote the complex spectral transform, it may be simplified to

$$x[t] = \int_{-\infty}^{\infty} X[f] \exp(i2\pi ft) \, df \ . \tag{5.2.16}$$

This expresses the relationship that

$x[t]$ is the Fourier transform of $X[f]$. $\qquad$ (5.2.17)

The analogous equation for $X[f]$ is

$$X[f] = \int_{-\infty}^{\infty} x[t] \exp(-i2\pi ft) \, dt \ , \tag{5.2.18}$$

i.e.

$X[f]$ is the Fourier transform of $x[t]$. $\qquad$ (5.2.19)

The two relationships together are known as a Fourier transform pair. Note that one must carry a negative exponent. In this text the negative is associated with $x[t]$ in the integral but it could have been used with $X[f]$ so long as it was used consistently.

5.3 Differentiation
Now that the empirical data may be represented by a continuous function it is a simple matter to calculate the derivative of that function. Using Equation (5.2.16), we find

$$\frac{d(x[t])}{dt} = \int_{-\infty}^{\infty} \frac{d}{dt} \{X[f] \exp(i2\pi ft)\} \, df \tag{5.3.1}$$

$$= \int_{-\infty}^{\infty} i2\pi f X[f] \exp(i2\pi ft) \, df \tag{5.3.2}$$

and, substituting for $X[f]$ and $\exp(i2\pi ft)$, we obtain

$$\frac{d(x[t])}{dt} = \int_{-\infty}^{\infty} i2\pi f(a[f] - ib[f])\{\cos(2\pi ft) + i\sin(2\pi ft)\} \, df \tag{5.3.3}$$

$$= \int_{-\infty}^{\infty} 2\pi f[\{b[f]\cos(2\pi ft) - a[f]\sin(2\pi ft)\}$$

$$+ i\{a[f]\cos(2\pi ft) + b[f]\sin(2\pi ft)\}] \, df \ . \tag{5.3.4}$$

But, if $d(x[t])/dt$ is real, the imaginary term may be dropped. In summation form then

$$\frac{d(x[t])}{dt}[j] = \sum_{k=0}^{n/2} \frac{2\pi k}{n} \left\{ b[k]\cos\left(\frac{2\pi kj}{n}\right) - a[k]\sin\left(\frac{2\pi kj}{n}\right) \right\} \ . \tag{5.3.5}$$

It should be clear from the plotted curve of Exercise 4.10.4 that the application of Equation (5.3.5) may give completely different results from those which may be obtained from finite differencing, even to the extent

of reversing the sign. This is because the first differential is directly
dependent upon frequency (factor $2\pi f$). Hence reliable slope estimates
may be obtained only in cases where the high frequencies have been
recorded. In turn this is dependent upon the frequencies present and the
datum spacing, since at least two points in a cycle are required for
determining its phase and amplitude (i.e. highest significant $k \leqslant T/2\Delta t$).

5.4 The convolution theorem

5.4.1 The transform of the product of two functions

If the function $x[t]$ is multiplied by another function $h[t]$, it may be
asked what the transform of the product looks like.

Consider the simple example of two symmetrical functions:

$$x_E[t] = 3\cos\left(\frac{2\pi 4t}{n\Delta t}\right) = \tfrac{3}{2}\exp\left(-\frac{i2\pi 4t}{n\Delta t}\right) + \tfrac{3}{2}\exp\left(\frac{i2\pi 4t}{n\Delta t}\right) \qquad (5.4.1.1)$$

and

$$h_E[t] = \tfrac{1}{2} + \tfrac{1}{4}\cos\left(\frac{2\pi t}{n\Delta t}\right) = \tfrac{1}{8}\exp\left(-\frac{i2\pi t}{n\Delta t}\right) + \tfrac{1}{2} + \tfrac{1}{8}\exp\left(\frac{i2\pi t}{n\Delta t}\right) . \qquad (5.4.1.2)$$

Then the product

$$x_E[t]h_E[t] = \tfrac{3}{2}\cos\left(\frac{2\pi 4t}{n\Delta t}\right) + \tfrac{3}{4}\cos\left(\frac{2\pi t}{n\Delta t}\right)\cos\left(\frac{2\pi 4t}{n\Delta t}\right) , \qquad (5.4.1.3)$$

which, with the use of the trigonometric relationship

$$2\cos R \cos S = \cos(R+S) + \cos(R-S) , \qquad (5.4.1.4)$$

becomes

$$x_E[t]h_E[t] = \tfrac{3}{8}\cos\left(\frac{2\pi 3t}{n\Delta t}\right) + \tfrac{3}{2}\cos\left(\frac{2\pi 4t}{n\Delta t}\right) + \tfrac{3}{8}\cos\left(\frac{2\pi 5t}{n\Delta t}\right) \qquad (5.4.1.5)$$

$$= \tfrac{3}{16}\exp\left(-\frac{i2\pi 5t}{n\Delta t}\right) + \tfrac{3}{4}\exp\left(-\frac{i2\pi 4t}{n\Delta t}\right) + \tfrac{3}{16}\exp\left(-\frac{i2\pi 3t}{n\Delta t}\right)$$

$$+ \tfrac{3}{16}\exp\left(\frac{i2\pi 3t}{n\Delta t}\right) + \tfrac{3}{4}\exp\left(\frac{i2\pi 4t}{n\Delta t}\right) + \tfrac{3}{16}\exp\left(\frac{i2\pi 5t}{n\Delta t}\right) ,$$

$$(5.4.1.6)$$

which is directly obtainable from the product of the exponentials.

The spectrum of $x[t]h[t]$ is known as the convolution of the spectra
of $x[t]$ and of $h[t]$. This may be written

$$\text{transform of } x[t]h[t] = X[f] * H[f] , \qquad (5.4.1.7)$$

where the operation on $X[f]$ and $H[f]$, denoted by an asterisk $*$, is called
a *convolution* (see Figure 5.4.1.1).

The mathematical expression for convolution is derived from the transform of the product $x[t]h[t]$.

$$\text{Transform of } x[t]h[t] = \int_{-\infty}^{\infty} x[t]h[t]\exp(-i2\pi f_1 t)dt ,\qquad (5.4.1.8)$$

but

$$x[t] = \int_{-\infty}^{\infty} X[f]\exp(i2\pi ft)df .$$

Therefore

transform of $x[t]h[t]$

$$= \int_{-\infty}^{\infty}\left\{\int_{-\infty}^{\infty} X[f]\exp(i2\pi ft)df\right\}h[t]\exp(-i2\pi f_1 t)dt \qquad (5.4.1.9)$$

$$= \int_{-\infty}^{\infty} X[f]\left\{\int_{-\infty}^{\infty} h[t]\exp[i2\pi(f_1-f)t]dt\right\}df \qquad (5.4.1.10)$$

$$= \int_{-\infty}^{\infty} X[f]H[f_1-f]df . \qquad (5.4.1.11)$$

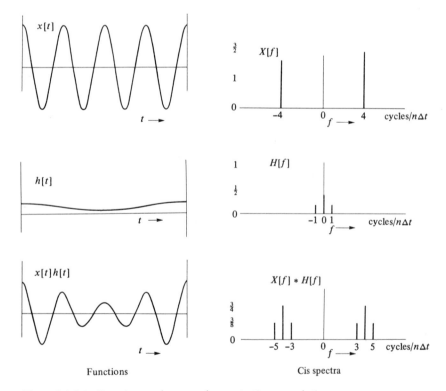

Functions   Cis spectra

Figure 5.4.1.1. Functions and spectra demonstrating convolution.

The change of order of integration between Equations (5.4.1.9) and (5.4.1.10) is justified by Fubini's theorem. The change may be demonstrated by the following example. Let

$$x[t] = \int_{-\infty}^{\infty} X[f] \exp(i2\pi ft)\,df$$

$$= (... + X[-2]\exp(-i2\pi 2t) + X[-1]\exp(-i2\pi t) + X[0]\exp(0)$$
$$+ X[1]\exp(i2\pi t) + ...)\Delta f. \qquad (5.4.1.12)$$

Then Equation (5.4.10) in summation form becomes equal to

$$\sum_t \{... + X[-2]\exp(-i2\pi 2t) + X[-1]\exp(-i2\pi t) + X[0]\exp(0)$$

$$+ X[1]\exp(i2\pi t) + ...\}h[t]\exp(-i2\pi f_1 t)\Delta f \Delta t \qquad (5.4.1.13)$$

$$= \{... + X[-2]\sum_t h[t]\exp[-i2\pi(f_1+2)t]$$

$$+ X[-1]\sum_t h[t]\exp[-i2\pi(f_1+1)t]$$

$$+ X[0]\sum_t h[t]\exp[-i2\pi(f_1-0)t]$$

$$+ X[1]\sum_t h[t]\exp[-i2\pi(f_1-1)t] + ...\}\Delta f \Delta t \qquad (5.4.1.14)$$

$$= \{... + X[-2]H[f_1+2] + X[-1]H[f_1+1]$$

$$+ X[0]H[f_1-0] + X[1]H[f_1-1] + ...\}\Delta f \qquad (5.4.1.15)$$

$$= \sum_f X[f]H[f_1-f]\Delta f \qquad (5.4.1.16)$$

$$= \int_{-\infty}^{\infty} X[f]H[f_1-f]\,df. \qquad (5.4.1.11)$$

In discrete terms, substituting for $X[f]$ and $H[f_1-f]$ separately using the relationship in Equation (5.2.14) and for df where $\Delta f = 1/n\Delta t$, Equation (5.4.1.11) becomes

$$\text{transform of } x[j]h[j] = \sum_{-\infty}^{\infty} X[k]\,n\Delta t\,H[k_1-k]\,n\Delta t\,\frac{1}{n\Delta t}. \qquad (5.4.1.17)$$

Δt is assumed to be unity. Thus the transform of the product of two discrete functions is the convolution of the two sets of Fourier coefficients [Equation (4.5.6)] divided by the frequency increment.

5.4.2 A numerical example of convolution

Let $X_E[k]$ and $H_E[k]$ be two-sided transforms of even functions with estimates as in Figure 5.4.2.1 given as follows.

k	-8	-7	-6	-5	-4	-3	-2	-1	0	1	2	3	4	5	6	7	8
$X_E[k]$	2	0	1	4	5	2	1	3	5	3	1	2	5	4	1	0	2
$nH_E[k]$	0	0	0	0	0	0	$\frac{1}{10}$	$\frac{1}{5}$	$\frac{2}{5}$	$\frac{1}{5}$	$\frac{1}{10}$	0	0	0	0	0	0

The calculations when $k_1 = -6$ are the following.

k	$X_E[k]$	$nH_E[-6-k]$	Product
-8	2	$\frac{1}{10}$	$\frac{2}{10}$
-7	0	$\frac{1}{5}$	0
-6	1	$\frac{2}{5}$	$\frac{2}{5}$
-5	4	$\frac{1}{5}$	$\frac{4}{5}$
-4	5	$\frac{1}{10}$	$\frac{5}{10}$

$$\sum X_E[k]\,nH_E[-6-k] = 1\tfrac{9}{10}$$

The calculations for all k_1's are as follows.

k_1	-6	-5	-4	-3	-2	-1	0
	$\frac{2}{10}$	0	$\frac{1}{10}$	$\frac{3}{10}$	$\frac{5}{10}$	$\frac{3}{10}$	$\frac{1}{10}$
	0	$\frac{1}{5}$	$\frac{4}{5}$	$\frac{1}{3}$	$\frac{3}{5}$	$\frac{3}{5}$	$\frac{3}{5}$
$X_E[k]\,nH_E[k_1-k]$	$\frac{2}{5}$	$\frac{8}{5}$	$\frac{10}{5}$	$\frac{4}{5}$	$\frac{2}{5}$	$\frac{6}{5}$	$\frac{10}{5}$
	$\frac{4}{5}$	$\frac{5}{5}$	$\frac{2}{5}$	$\frac{5}{5}$	$\frac{2}{5}$	$\frac{1}{5}$	$\frac{3}{5}$
	$\frac{5}{10}$	$\frac{2}{10}$	$\frac{1}{10}$	$\frac{4}{10}$	$\frac{5}{10}$	$\frac{2}{10}$	$\frac{1}{10}$
$\sum X_E[k]\,nH_E[k_1-k] =$	$1\tfrac{9}{10}$	3	$3\tfrac{2}{5}$	$2\tfrac{7}{10}$	$2\tfrac{2}{5}$	$2\tfrac{9}{10}$	$3\tfrac{2}{5}$

Because the functions $X_E[k]$ and $H_E[k]$ are even, the convolution at positive frequencies gives the same results as at negative frequencies. All are plotted in Figure 5.4.2.1.

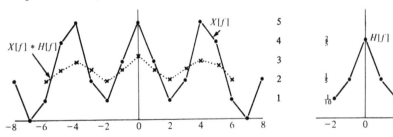

Figure 5.4.2.1.

5.4.3 General relationships between spectra and products

It can be seen that the spectrum of one function, say $x[t]$, is modified by $h[t]$ such that it is spread out into the form of the spectrum of $h[t]$. When the spectra are continuous this process tends to blur the final spectrum, since the spreading out occurs at every possible frequency of $x[t]$. Therefore any particular frequency is the sum of all weighted amplitudes resulting from the spreading out of neighboring frequencies.

It is now possible to write down the general relationships for products corresponding to equations:

$$\int_{-\infty}^{\infty} X[f]H[f_1 - f]\,df \text{ is the Fourier transform of } x[t]h[t] \ , \qquad (5.4.3.1)$$

$$X[f]H[f] \text{ is the Fourier transform of } \int_{-\infty}^{\infty} x[t]h[t_1 - t]\,dt \ , \qquad (5.4.3.2)$$

and

$$X[f]H^*[f] \text{ is the Fourier transform of } \int_{-\infty}^{\infty} x[t_1 + t]h[t]\,dt \ , \qquad (5.4.3.3)$$

where the superior asterisk * signifies the conjugate function, obtained by changing the sign of i.

For functions which contain odd and even parts the product on the left-hand side of Equation (5.4.3.3), for example, is given by

$$\begin{aligned}
X[f]H^*[f] &= (a_x[f] - ib_x[f])(a_h[f] + ib_h[f]) \\
&= (a_x[f]a_h[f] + b_x[f]b_h[f]) + i(a_x[f]b_h[f] - a_h[f]b_x[f]) \ ,
\end{aligned}$$
$$(5.4.3.4)$$

where the real part is even and the imaginary part (associated with i) is odd.

In discrete terms, a substitution is made for each density and then the equation is multiplied by the frequency increment $1/n$, so that it applies to *two*-sided discrete bands

$$X[k]\mathcal{H}H^*[k]n\frac{1}{\mathcal{n}} = (a_x[k]a_h[k] + b_x[k]b_h[k])\frac{n}{4}$$

$$+ i(a_x[k]b_h[k] - a_h[k]b_x[k])\frac{n}{4} \ , \qquad (5.4.3.5)$$

where the a's and b's are calculated by Equations (3.3.11) and (3.3.12).

Other names for convolution are the following: superposition theorem, Faltung's integral, Green's theorem, Duhamel's theorem, Borel's theorem, and Boltzmann–Hopkinson's theorem.

It should be apparent that convolution is not a simple operation if the shorter function is a complicated one. On the other hand, if both functions are transformed, only a direct multiplication is required. This is one reason for the importance of the Fourier transform in pattern recognition.

5.5 Exercises

5.5.1 Plot the one-sided $a[f]$ density spectrum of the data in Exercise 1.6.1. Label and scale the axes.

5.5.2 Calculate the first differences of the same data using the convolution theorem. Compare with actual first differences. Note that with large magnitudes and large amounts of data there may be 'round-off' error.

5.5.3 Calculate the first differential of the same data and compare with the results of Exercise 5.4.2 and the plot of Exercise 4.10.4.

5.5.4 Write down the integral equation for $x[t]$ in similar format to Equation (5.3.5).

5.5.5 Reading: Barber (1961); Jennison (1961, particularly chapter 6); Cooley *et al.* (1967b).

6

Filtering

6.1 Introduction

As will be seen later, filtering is important in spectral analysis as its application helps remove some bias from the estimates. However, it does have wider application. For example, in geography, averages of data are often calculated before further analysis is conducted. Averaging, one form of filtering, alters the data, yet little consideration is given to the effects of such a statistical procedure. If the data are arranged in some form of sequence (or array), then averaging will remove some detailed information and leave the broad-scale fluctuations almost untouched, i.e. the high frequencies are filtered out.

What is the influence, with respect to scale, of averaging? Is the straight average what is really required? An attempt is made here to answer the first of these questions and on the basis of the results from this section the user may decide for himself what kind of filter he needs. An extremely good survey of the field is given by Holloway (1958) and only some of the points will be repeated here.

Filtering is just a general name given to the process of systematically modifying data which are arranged in a sequence or array.

Averaging is one form of *smoothing*, which in turn is one form of filtering. It removes the high- and, in part, the medium-scale disturbances (frequencies) from the data. In other words, it is a *low-pass filter* since it allows the low frequencies through. At the other extreme there are *high-pass filters* and, between, *band-pass filters*.

As a slight variation of filtering, instead of the removal of frequencies to leave a band untouched, this band may be amplified or *pre-emphasised* above the others. Such a procedure is followed at a radio broadcasting station. The audio frequencies are amplified to such an extent that interference from other amplitudes introduced at various stages along the transmission is relatively small. At the receiving station (radio) the pre-emphasised amplitudes are reduced through a process known as *equalisation*. Since the interference (noise) amplitudes are reduced proportionately, only those sounds originally broadcast will be heard. The original balance may be returned to a smoothed series in the same way. This is known as *inverse smoothing*.

6.2 The effect of filtering

Filtering is performed with a filtering function which, in the practical case here, will consist of a series of discrete weights, i.e. ($\Delta j = 1$)

$$x[j_1] = \sum_{j=-u}^{v} x[j_1+j]w[j]\,\Delta j \quad \text{[Equation (4.1) of Holloway (1958)]}$$
$$\text{(6.2.1)}$$

$$= x[j_1-u]w[-u] + ... + x[j_1]w[0] + ... + x[j_1+v]w[v] \quad \text{(6.2.2)}$$

(see Figure 6.2.1). This is the right-hand side of Equation (5.4.3.3) for the continuous function

$$= \int_{-\infty}^{\infty} x[t_1+t]w[t]\,dt \quad \text{(6.2.3)}$$

where $w[t] = 0$ when $u > t > v$.

The simplest example of a filtering or a smoothing function is the running mean where

$$w[t] = w[-u] = w[-u+1] = ... = w[0] = ... = w[v] = \frac{1}{I} \quad \text{(6.2.4)}$$

where

$$I = |u| + |v| + 1 .$$

The effect of filtering on the original series may be judged by the proportionate change in the various parameters. For example, the effect

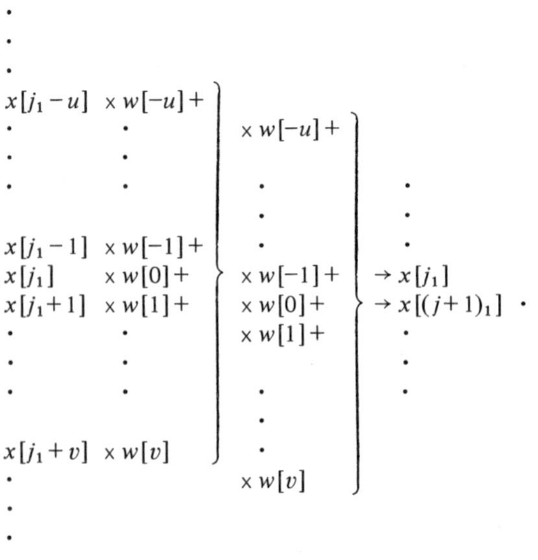

Figure 6.2.1.

upon the mean is given by the ratio of the output mean to that of the input mean.

$$\text{Ratio} = \frac{\text{mean out}}{\text{mean in}} = \frac{\frac{1}{n\Delta t}\int_{t_1}\int_t w[t]x[t_1+t]\,dt\,dt_1}{\frac{1}{n\Delta t}\int_t x[t]\,dt} \tag{6.2.5}$$

$$= \frac{\frac{1}{n\Delta t}\int_t x[t]\,dt \int_t w[t]\,dt}{\frac{1}{n\Delta t}\int_t x[t]\,dt} \tag{6.2.6}$$

$$= \int_{-\infty}^{\infty} w[t]\,dt \ . \tag{6.2.7}$$

Usually the mean is kept constant. In other words the ratio must be equal to one or from Equation (6.2.7) the sum of the weights must equal one.

The step from Equation (6.2.5) to (6.2.6) may be demonstrated by the following example:

weights $\quad w[-2] \quad w[-1] \quad w[0] \quad w[1] \quad w[2]$

series $\quad x[-\tfrac{1}{2}n] \ ... \ x[-1] \quad x[0] \quad x[1] \ ... \ x[\tfrac{1}{2}n]$.

The numerator in Equation (6.2.5) may be written $\sum_{t_1}\left(\sum_t w[t]x[t_1+t]\right)$,
where the brackets represent the new series in t_1 and may be specified ($\Delta t = \Delta t_1 = 1$) by

$x[(-\tfrac{1}{2}n+2)_1] = w[-2]x[-\tfrac{1}{2}n] \ +w[-1]x[-\tfrac{1}{2}n+1]+w[0]x[-\tfrac{1}{2}n+2]+w[1]x[-\tfrac{1}{2}n+3]+w[2]x[-\tfrac{1}{2}n+4]$

$$\vdots \qquad \vdots \qquad \vdots \qquad \vdots \qquad \vdots$$

$x[-1_1] \quad = w[-2]x[-3] \ +w[-1]x[-2] \ +w[0]x[-1] \ +w[1]x[0] \ +w[2]x[1]$

$x[0_1] \quad = w[-2]x[-2] \ +w[-1]x[-1] \ +w[0]x[0] \ +w[1]x[1] \ +w[2]x[2]$

$x[1_1] \quad = w[-2]x[-1] \ +w[-1]x[0] \ +w[0]x[1] \ +w[1]x[2] \ +w[2]x[3]$

$$\vdots \qquad \vdots \qquad \vdots \qquad \vdots \qquad \vdots$$

$x[(+\tfrac{1}{2}n-2)_1] = w[-2]x[\tfrac{1}{2}n-4]+w[-1]x[\tfrac{1}{2}n-3] \ +w[0]x[\tfrac{1}{2}n-2] \ +w[1]x[\tfrac{1}{2}n-1] \ +w[2]x[\tfrac{1}{2}n]$.

$$\tag{6.2.8}$$

The numerator now equals the sum of all terms on left, or right of Equation (6.2.8). This sum may be written as the sums of columns in Equation (6.2.8):

$$w[-2]\left(\sum x[t]\right) + w[-1]\left(\sum x[t]\right) + w[0]\left(\sum x[t]\right)$$

$$+ w[1]\left(\sum x[t]\right) + w[2]\left(\sum x[t]\right) \ . \tag{6.2.9}$$

It is assumed that adjustments may be made for the end point variation between the $\left(\sum x[t]\right)$'s and that the summation still covers the interval $-\tfrac{1}{2}n$ to $\tfrac{1}{2}n$. Therefore

$$\int_{t_1}\left(\int_t w[t]x[t_1+t]\,dt\right)dt_1 = \int_t x[t]\,dt \int_t w[t]\,dt \ . \tag{6.2.10}$$

In the same way, the variation between the original and the new series at a particular frequency may be assessed. The ratio is known as the frequency response:

$$\text{ratio } [f] = \text{response } [f] = \frac{\text{Fourier transform of new series at } [f]}{\text{Fourier transform of old series at } [f]}$$

$$= \frac{\text{Fourier transform of } \displaystyle\int_{-\infty}^{\infty} w[t]x[t_1+t]\,dt}{\text{Fourier transform of } x[t]}$$

$$\tag{6.2.11}$$

but because of Equation (5.4.3.3) it equals

$$\frac{X[f]\,W^*[f]}{X[f]} \tag{6.2.12}$$

$$= \int_{-\infty}^{\infty} w[t]\exp(i2\pi ft)\,dt \tag{6.2.13}$$

$$= \underbrace{\int_{-\infty}^{\infty} w[t]\cos(2\pi ft)\,dt}_{\text{real}} + i\underbrace{\int_{-\infty}^{\infty} w[t]\sin(2\pi ft)\,dt}_{\text{imaginary}} \ . \tag{6.2.14}$$

Therefore the magnitude of the response at frequency f is given by

$$|W^*[f]| = [[(\text{real, } W[f])^2 + (\text{imaginary, } W[f])^2]]^{\frac{1}{2}} \tag{6.2.15}$$

and the change of phase in new series relative to old by

$$\Phi[f] = \arctan\left(\frac{\text{imaginary, } W[f]}{\text{real, } W[f]}\right) \ . \tag{6.2.16}$$

If the filtering function is symmetrical, only the real response function has magnitude. Therefore there is no change of phase with a symmetrical function. Further, since a change of phase is usually undesirable, most mathematical weighting functions are symmetrical.

In summary, for the calculation of the response function for a symmetrical filter, i.e. the effect of the filter on some data, Equation (5.2.12) is applied with $w[t]$ replacing $x[t]$. In other words, the response function for a symmetrical filter is the cosine transform of that filter.

From Equations (5.4.3.3) and (6.2.11) it will be seen that a series modified by a filter before transformation may have its original frequencies restored through division by the response function $W^*[f]$, i.e.

$$X[f] = \frac{\int_{-\infty}^{\infty}\left(\int_{-\infty}^{\infty} w[t]x[t_1+t]\,dt\right)\exp(-i2\pi ft)\,dt}{W^*[f]} \qquad (6.2.17)$$

6.3 Examples of filtering functions

6.3.1 The constant

It should be clear by now that a constant in the time or space domain transforms to a single spike in the frequency domain. In other words, only $a[0]$ exists (Figure 6.3.1.1).

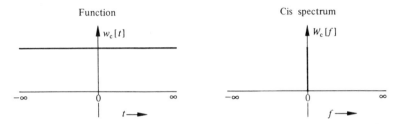

Figure 6.3.1.1.

6.3.2 The comb

The comb or series of equally spaced spikes, often known as the Dirac function or Delta function, is of utmost importance in digital computation since, through its use, a continuous record may be converted into equispaced observations. The comb of spikes Δt apart may be defined as

$$W_D[t] = \sum_{j=-\infty}^{\infty} \delta[t-j\Delta t]\,\Delta t = 1 \text{ for } |t| = j\Delta t \;, \qquad (6.3.2.1)$$

where $\delta[t-j\Delta t]$ is zero for $|t| \neq j\Delta t$. Its transform is given by

$$W_D[f] = \sum_{j=-\infty}^{\infty} \delta\left[f-\frac{j}{\Delta t}\right] . \qquad (6.3.2.2)$$

Therefore an infinite series of spikes Δt apart transforms into another infinite series of spikes $1/\Delta t$ apart as seen in Figure 6.3.2.1. Each spike

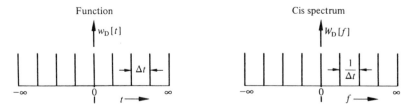

Figure 6.3.2.1.

of this periodic function in the frequency domain is the sum of all amplitude densities at $1/\Delta t$ apart. This is called aliasing and will be discussed in Section 6.4.

6.3.3 The rectangle

This function may be defined as

$$w[t] = \begin{cases} \dfrac{1}{I} & |t| \leqslant \dfrac{I}{2} \\[2mm] 0 & |t| > \dfrac{I}{2} \end{cases} , \tag{6.3.3.1}$$

where I is the averaging interval or number of terms over which the average is calculated.

This is an even function so only cosine terms are required:

$$W[f] = \int_{-I/2}^{I/2} \frac{1}{I} \cos(2\pi ft)\,dt = \frac{\sin(2\pi fI/2)}{2\pi fI/2} \tag{6.3.3.2}$$

or, if $w_0[t] = 1$, $|t| \leqslant I/2$,

$$W_0[f] = 2\frac{(I/2)\sin(2\pi fI/2)}{2\pi fI/2} . \tag{6.3.3.3}$$

The function and resulting cis spectrum is given in Figure 6.3.3.1.

When the function is in the form of equispaced data the result is slightly different. For the running mean of $I = 2M+1$ terms,

$$W_{0D}[f] = \frac{1}{I} \sum_{j=-M}^{M} \exp(-i2\pi fj\Delta t) . \tag{6.3.3.4}$$

If this sum is multiplied by $\sin(\pi f\Delta t) = \frac{1}{2}\{\exp(-i\pi f\Delta t) - \exp(i\pi f\Delta t)\}$ it will be seen that terms containing $j \neq M$ cancel out and

$$W_{0D}[f] = \frac{1}{I} \frac{\sin(\pi fI\Delta t)}{\sin(\pi f\Delta t)} , \tag{6.3.3.5}$$

which for small f is similar to Equation (6.3.3.2).

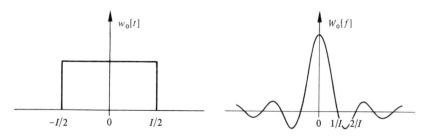

Figure 6.3.3.1.

6.3.4 The triangle

Defined as

$$
w[t] = \begin{cases} 1 - \dfrac{|t|}{I/2} & |t| \leqslant \dfrac{I}{2} \\[2ex] 0 & |t| > \dfrac{I}{2} \ ; \end{cases}
$$

(6.3.4.1)

this transforms, by use of integration by parts, to

$$
W[f] = \frac{I}{2}\left\{ \frac{\sin(\pi f I/2)}{\pi f I/2} \right\}^2 .
$$

(6.3.4.2)

Alternatively this result may be obtained by convolving $w_0[t]$ with itself. Therefore the transform may be squared, $W_0^2[f]$. From Equation (6.3.3.3),

$$
W_0^2[f] = 4\left\{ \frac{(I/2)\sin(2\pi f I/2)}{2\pi f I/2} \right\}^2 .
$$

(6.3.4.3)

This is a triangle with base $2I$ and height at apex of I and is defined by

$$
w_1[t] = I\left(1 - \frac{|t|}{I}\right) \qquad \text{for } |t| \leqslant I
$$

(6.3.4.4)

and Equation (6.3.4.2) becomes

$$
W_1[f] = I^2\left\{ \frac{\sin(\pi f I)}{\pi f I} \right\}^2
$$

(6.3.4.5)

which is the same as Equation (6.3.4.3).

For equi-spaced data Equation (6.3.4.5) becomes

$$
W_{1D}[f] = \left\{ \frac{\sin(\pi f M \Delta t)}{\sin(\pi f \Delta t)} \right\}^2 .
$$

(6.3.4.6)

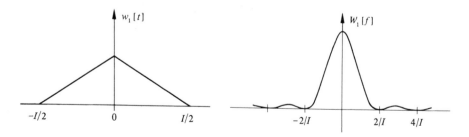

Figure 6.3.4.1.

6.3.5 The Cosine Bell

$$w_2[t] = \tfrac{1}{2}\left\{1 + \cos\left(\frac{2\pi|t|}{I}\right)\right\} \qquad |t| \leqslant \frac{I}{2} \qquad (6.3.5.1)$$

$$= 0 \qquad\qquad |t| > \frac{I}{2} \;.$$

Then

$$W_2[f] = \tfrac{1}{4}W_0\left[f - \frac{1}{I}\right] + \tfrac{1}{2}W_0[f] + \tfrac{1}{4}W_0\left[f + \frac{1}{I}\right] \;. \qquad (6.3.5.2)$$

Similarly for equi-spaced data Equation (6.3.5.2) becomes

$$W_{2D}[f] = \tfrac{1}{4}W_{0D}\left[f - \frac{1}{2M\Delta t}\right] + \tfrac{1}{2}W_{0D}[f] + \tfrac{1}{4}W_{0D}\left[f + \frac{1}{2M\Delta t}\right] \;. \qquad (6.3.5.3)$$

Other examples may be found, for example, in Barber (1961, 1966), Blackman and Tukey (1958), Holloway (1958), and Jennison (1961).

6.4 Aliasing

In practice a continuous function, $x[t]$, is often represented by a series of equispaced observations, $x[j]$. Such a series may be thought of as being produced by multiplication with a function which is composed of ones and intervening zeros (a Dirac comb). The resulting spectrum therefore is the convolution of the spectra of the original function and the comb. If the end values of the comb are assumed to be $\tfrac{1}{2}$, its spectrum is given by [see Equation (6.3.3.4) for derivation]

$$W_D[f] = 2M\Delta t \cos(\pi f\Delta t)\frac{\sin(2\pi fM\Delta t)/2\pi fM\Delta t}{\sin(\pi f\Delta t)/\pi f\Delta t} \;. \qquad (6.4.1)$$

This function is periodic at intervals of integer multiples of f as shown in Figure 6.4.1 for eleven observations ($M = 5$, $\Delta t = 1$). Furthermore, it will be noted that $W_D[f]$ is zero at all other calculable frequencies in discrete analysis. Consequently each estimate in the final spectrum of $x[j]$ will be the sum of the estimates at all the periodic frequencies of $x[t]$.

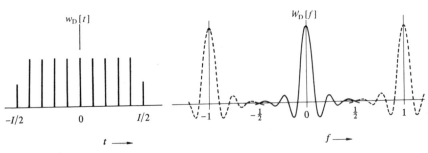

Figure 6.4.1.

The latter statement may be expressed in a different way. The use of discrete data points introduces a cut-off frequency in the spectrum at $k = n/2$, or $f_F = \frac{1}{2}$ (when $\Delta t = 1$), the Nyquist frequency. Since all the variance in the data must be accounted for, it must be distributed amongst the bands below f_F. However, some of this variance may come from the higher unresolvable frequencies.

For example, in Figure 6.4.2 two curves of differing frequencies have been fitted to the same data. Mathematically only the continuous curve may be resolved, and the variance must be attributed to that, yet the data may be generated completely by the dashed curve.

This folding in of frequencies greater than f_F is known as aliasing and the frequencies between 0 and f_F known as the principal aliases.

The positive aliased frequencies of the principal alias of f are

$$2f_F - f, \quad 2f_F + f, \quad 4f_F - f, \quad 4f_F + f, \quad \text{etc.} \tag{6.4.2}$$

For example, if $\Delta t = 1$ day, the first aliased frequency of $\frac{1}{5}$ cycles per day (period of 5 days) is given by

$$2f_F - f = 2\frac{1}{2 \times 1} - \frac{1}{5} = \frac{4}{5} \text{ cycles per day} \tag{6.4.3}$$

or

$$= 1\frac{1}{4} \text{ days' period.}$$

The calculated aliased spectrum $-f_F$ to f_F may be expressed as the sum of true spectral densities:

$$X_L[f] = X[f] + X[-f] + X[2f_F - f] + X[-(2f_F - f)] + \text{etc.} \tag{6.4.4}$$

If $X[f]$ is symmetrical, this becomes

$$2X_L[f] = 2X[f] + 2X[2f_F - f] + 2X[2f_F + f] + \text{etc.} \tag{6.4.5}$$

Nearly always some aliasing will occur. The student must judge for himself whether much variance lies in the higher frequencies and whether such variance will invalidate any estimates he makes of the variance spectrum in the lower frequencies (see, for example, Figure 6.4.3).

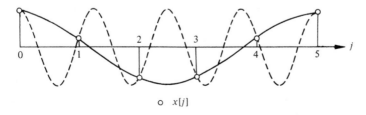

$\circ \quad x[j]$

Figure 6.4.2.

Not all the variance present in the higher frequencies will be incorporated into the widely spaced data points; so the folding back will not be as complete as shown in Figure 6.4.3. However, it would be sufficient to create serious variations from the 'true' spectrum.

Aliasing may be reduced only through increasing f_F to a point beyond which the variance is insignificant. This may be done by decreasing Δt. If f_F cannot be made sufficiently large, then there is no point attempting spectral analysis.

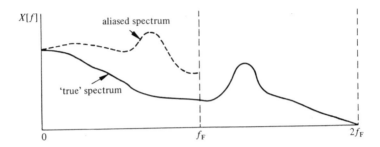

Figure 6.4.3.

6.5 Exercises

6.5.1 If a five-point non-overlapping running mean is applied to 100 periodic observations, what is the effect upon each amplitude (i.e. response, $W^*[k]$) for all calculable k?

6.5.2 What would be the phase change at each calculable frequency if the following non-overlapping function were applied to 100 periodic observations: $0 \cdot 05, 0 \cdot 2, 0 \cdot 3, 0 \cdot 4, 0 \cdot 05$?

6.5.3 Calculate the resulting twelve points after applying a perfect filter, which passes all amplitudes for frequencies $0 \leqslant k \leqslant 2$ and passes none for $k > 2$, to the periodic data in Exercise 1.6.1.

6.5.4 What are the magnitudes of the weights of the perfect filter in Exercise 6.5.3?

6.5.5 Reading: Blackman and Tukey (1958, especially Section A12 and Appendix A); Holloway (1958); Tobler (1969a, 1969b).

Spectral analysis of non-periodic functions

7.1 Introduction to the analysis of sequenced data

It is now necessary to consider some assumptions upon which the statistical analyses of sequenced data are based. Bendat and Piersol (1966) give a good introduction to the subject in their first chapter, so only a few remarks will be made here.

The immediate problem in the statistical analysis of data, say $x[t]$, is the specification of the population parameters from which it came. In simple problems of a single variable where observations are physically independent and normal, the mean and variance characterise the distribution completely. If $x[t]$ has zero mean (i.e. mean removed), then

$$E(x[t]) = \mu = \lim_{T \to \infty} \frac{1}{T} \int_{-T/2}^{T/2} x[t]\,dt = 0 \tag{7.1.1}$$

$$E(x^2[t]) = \sigma^2 = \lim_{T \to \infty} \frac{1}{T} \int_{-T/2}^{T/2} x^2[t]\,dt \ . \tag{7.1.2}$$

However, as pointed out in Section 1.3.2, sequenced data are frequently not independent and at least another parameter, the autocovariance function, is required.

$$E(x[t]x[t+p]) = xx[p] = \lim_{T \to \infty} \frac{1}{T} \int_{-T/2}^{T/2} x[t]x[t+p]\,dt \ . \tag{7.1.3}$$

Alternatively this may be converted into a probability density function if Equation (7.1.3) is normalised by dividing through by $xx[0]$, the variance.

For random data $xx[p]$ looks like Figure 7.1.1. Except for the term at $p = 0$ there is, on the average, no correlation between terms p distance apart.

These three terms μ, σ^2, $xx[p]$ are, from a statistical point of view, ensemble parameters. Consequently, if only one spatial or temporal

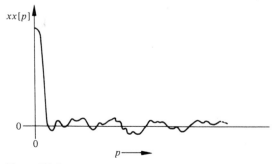

Figure 7.1.1.

sequence $x[t]$ is to be used for their estimation, some assumptions must be made about the relationship of the sequence to the ensemble. In fact the sequence may be thought of as a subset of an ensemble.

For example, let there be a sample of N sequences in the ensemble with J referring to a particular sequence known as a sample function. In each sequence let there be a sample of n observations known as a sample record with j referring to a particular observation (see Figure 7.1.2).

Two types of mean and autocovariance functions may now be defined. For the ensemble ($\Delta J = 1$)

$$\text{mean } J = \lim_{N \to \infty} \frac{1}{N \Delta J} \int_J x_J[t] \, dJ \tag{7.1.4}$$

$$= \lim_{N \to \infty} \frac{1}{N} \sum_J x_J[j] \quad , \tag{7.1.5}$$

$$\text{cov } J = \lim_{N \to \infty} \frac{1}{N \Delta J} \int_J x_J[t] x_J[t+p] \, dJ \tag{7.1.6}$$

$$= \lim_{N \to \infty} \frac{1}{N} \sum_J x_J[j] x_J[j+p] \quad . \tag{7.1.7}$$

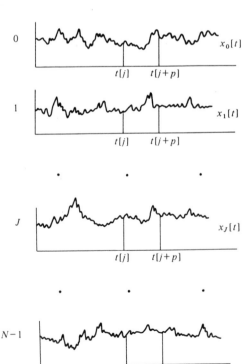

Figure 7.1.2.

For a particular sequence ($\Delta j = 1$)

$$\text{mean } j = \lim_{n \to \infty} \frac{1}{n} \sum_j x_J[j] \; , \tag{7.1.8}$$

$$\text{cov } j = \lim_{n \to \infty} \frac{1}{n} \sum_j x_J[j] x_J[j+p] \; . \tag{7.1.9}$$

If mean J and cov J do not vary significantly as j changes, the ensemble is said to be weakly statistically stationary. If the higher moments are also constant, then the ensemble is strongly stationary.

If mean j and cov j do not vary significantly as $x_J[j]$ changes, then the ensemble is said to be ergodic. It should be noted that ergodicity presupposes stationarity.

Since only a sample is available, the mean J, cov J, mean j, and cov j should be written (mêan J), (côv J), (mêan j), and (côv j). Even though stationarity and ergodicity apply, these sample estimates will vary according to their statistical sampling distributions.

If only one sample function is available, the term stationary is often used in reference to different sample records from that function. To distinguish this type of stationarity the phrase 'self-stationary' is used. Thus useful assumptions for estimating the population parameters from a sequence are normality and stationarity. Goodman (1961) has pointed out that time series are usually non-ergodic and that the assumption of ergodicity is not necessary, yet Blackman and Tukey (1958) use it.

Even if $xx[p]$ is a sufficient descriptor, it is difficult to interpret, since adjacent estimates are not independent and the confidence intervals are difficult to calculate. On the other hand, the Fourier transform of this function is relatively simple and its statistical estimates relatively stable and independent. Consequently, Tukey (1949) and others have advocated the use of $XX[f]$, which turns out to be the spectrum of the variance of $x[t]$ as shown below.

$$xx[p] = \lim_{T \to \infty} \frac{1}{T} \int_{-\infty}^{\infty} x[t] x[t+p] \, dt \; . \tag{7.1.10}$$

If $h[t]$ is put equal to $x[t]$ in Equation (5.4.3.3), the spectrum (Fourier transform) of this function may be written

$$XX[f] = \int_{-\infty}^{\infty} \left[\lim_{T \to \infty} \frac{1}{T} \int_{-\infty}^{\infty} x[t] x[t+p] \, dt \right] \exp(-i2\pi fp) \, dp \tag{7.1.11}$$

$$= \lim_{T \to \infty} \frac{1}{T} X[f] X^*[f] \tag{7.1.12}$$

$$= \lim_{T \to \infty} \frac{1}{T} |X[f]|^2 \tag{7.1.13}$$

but through Equation (5.2.18) it equals

$$\lim_{T \to \infty} \frac{1}{T} \left| \int_{-\infty}^{\infty} x[t] \exp(-i2\pi ft)\,dt \right|^2 . \tag{7.1.14}$$

Since $x[t]$ is zero outside the interval $-T/2$ to $T/2$, this may be written

$$XX[f] = \lim_{T \to \infty} \frac{1}{T} \left| \int_{-T/2}^{T/2} x[t] \exp(-i2\pi ft)\,dt \right|^2 , \tag{7.1.15}$$

which is the spectrum of the variance, and also

$$XX[f] = \int_{-\infty}^{\infty} xx[p] \exp(-i2\pi pf)\,dp . \tag{7.1.16}$$

The relationship between the variance spectrum and the autocovariance is known as the Wiener–Khintchine relation.

The spectrum corresponding to Figure 7.1.1, random data, sometimes known as white noise, is given in Figure 7.1.3.

Of course higher moments may be considered in the same way. Hence the terms bispectra and polyspectra are now appearing in the literature.

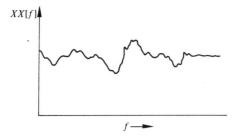

Figure 7.1.3.

7.2 The real relationship between the autocovariance function and the variance spectrum

In reality Equation (7.1.16) cannot be applied, as $xx[p]$ is not defined outside the limits of p. This follows from the fact that only a finite length of $x[t]$ is available. In order to develop Equation (7.1.16) it was assumed that $x[t]$ was zero for $|t| > T/2$, but this is usually not the case. Also, not only is p limited but $xx'[p]$ is obtained from a varying length of record, i.e.

$$xx'[p] = \frac{1}{(n-|p|)\Delta t} \int_{-\{(n-|p|)/2\}\Delta t}^{\{(n-|p|)/2\}\Delta t} x\left[t - \frac{P}{2}\right] x\left[t + \frac{P}{2}\right] dt \tag{7.2.1}$$

$$= \frac{1}{n-|p|} \sum_{j=0}^{(n-1-|p|)} x[j] x[j+p] . \tag{7.2.2}$$

Now in the range of p ($-m$ to m), $xx'[p]$ is an estimate of the ensemble value of $xx[p]$ and

$$\text{ensemble average } xx'[p] = xx[p] \qquad |p| \leqslant m \ . \tag{7.2.3}$$

Here Blackman and Tukey (1958) invoke ergodicity as mentioned in the last section.

In order to make Equation (7.2.3) apply to all values of p, each side is multiplied by a function $h[p]$, which is one at $p = 0$ and zero at $|p| > m$ (Figure 7.2.1). Elsewhere $h[p]$ has yet to be specified.

Then

$$\hat{xx}[p] = xx'[p]h[p] \tag{7.2.4}$$

and Equation (7.2.3) becomes

$$\text{ave } \hat{xx}[p] = \text{ave } xx'[p]h[p] = xx[p]h[p] \ , \tag{7.2.5}$$

where 'ave' means ensemble average.

The transform of this equation is

$$\text{ave } \hat{XX}[f] = \text{ave } XX'[f] * H[f] = XX[f] * H[f] \ , \tag{7.2.6}$$

where $XX[f]$ is the true spectrum of the variance of $x[t]$.

Theoretically $xx'[p]$ cannot be transformed since it is undefined for $|p| > m$ but $\hat{xx}[p]$ has no such restriction and its transform is the estimate for $XX[f]$. Clearly it is a blurred (smoothed over frequency) estimate, the degree of blurring depending upon the function $h[p]$ since, from Equation (5.4.3.1), Equation (7.2.6) becomes

$$\text{ave } \hat{XX}[f_1] = \int_{-\infty}^{\infty} XX[f]H[f_1 - f]\,df \ . \tag{7.2.7}$$

The blurring is perhaps more clearly demonstrated by the situation where $xx'[p]$ is known exactly for $|p| \leqslant m$, i.e.

$$xx'[p] = xx[p] \text{ for } |p| \leqslant m \ . \tag{7.2.8}$$

Then

$$\hat{xx}[p] = xx[p]h[p]$$

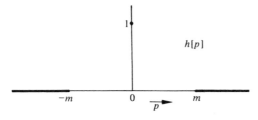

Figure 7.2.1.

where $h[p]$ is the rectangle of Section 6.3.3, i.e.

$$h[p] = 1 \qquad |p| \leqslant m$$
$$ = 0 \qquad |p| > m . \tag{7.2.9}$$

It can be thought of as a slit or window through which $xx[p]$ is viewed. Consequently Blackman and Tukey (1958) have called $h[p]$, whatever its form, a lag window. The corresponding transformed function $H[f]$ is therefore known as a spectral window but the word 'kernel' is frequently used. If $h[p]$ produces discrete data, it will also lead to aliasing.

To return to Equation (7.2.9) this is called a 'do nothing' window by Blackman and Tukey since its application is the equivalent of transforming $xx'[p]$ for $|p| \leqslant m$. It is a poor window since it has large secondary lobes in the spectral domain (Figure 6.3.3.1). These are both negative and positive so their effect may smother completely the true spectral density. (See Section 5.4.2 for the way a particular $H[f]$ affects $XX[f]$.)

Obviously better windows are available and much effort has gone into developing spectral windows which concentrate the averaging around the estimating frequency. However, it will be seen in Section 7.6 that the more concentrated is the window the less stable are the results (i.e. confidence bands become wide). At the other extreme wide windows and narrow confidence bands tend to produce bias through averaging out real peaks and valleys (Jenkins and Watts, 1968, p.247). Generally it is better initially to take some intermediate window which has most weight in a broad main lobe and little in the side lobe. The most frequently used windows are given in Figure 1 of Blackman and Tukey (1958) and Figure 2 of Jenkins (1961). The Tukey cosine bell which will be used here is given in Section 6.3.5.

In summary, one of the best estimates available for the true spectrum will be the transform of the filtered sample autocovariance function given by Equations (7.2.4) and (7.2.2).

7.3 Practical steps using the autocovariances, the indirect method

The basic equations may now be written down for the practical case of a series of n equally spaced observations. Both the mean and trend components should be initially removed or they will appear at zero frequency and blur the function at higher frequencies. The lag window which must be of the comb type need not be applied to the autocovariance. Indeed it is often simpler to introduce the corresponding spectral window at the end when convolution turns out to be a simple set of calculations.

The major steps therefore are as follows:

Step 1 remove mean and trends.

Step 2 calculate the sample autocovariances as given by Equation (7.2.2):

$$xx'[p] = \frac{1}{n - |p|} \sum_{j=0}^{(n-1-|p|)} x[j]\,x[j+p] . \tag{7.3.1}$$

Step 3 transform the sample autocovariance function to give $XX'[f]$ which is an even function:

$$XX'[f] = \sum_{p=-m}^{m} xx'[p]\cos(2\pi fp)\Delta p \ , \qquad (7.3.2)$$

where $\Delta p =$ constant $\times \Delta t$, the lag increment. The frequency f, which in Equation (5.2.11) was defined as $k/n\Delta t$, here corresponds to $r/2m\Delta p$, where $2m\Delta p$ ls the length of the basic interval for the autocovariance function and r is an integer varying between $-m$ and m. Since $XX'[f]$ is an even function, only estimates corresponding to positive values of r are calculated then doubled. Further, if the summation goes from 0 to m, the final estimates must be doubled. Combined, these give rise to a multiplication by 4. If the variance density estimates are converted into average variance per frequency band, Equation (7.3.2) must be multiplied by $\frac{1}{2}m\Delta p$ (see Section 5.2).

$$XX'[r] = \frac{4\Delta p}{2m\Delta p} \sum_{p=0}^{m} xx'[p]\cos(2\pi fp) \ . \qquad (7.3.3)$$

Adjustments must be made for the end data, so that the explicit equations for the estimates, known as line powers, may be written

$$XX'[0] = \frac{1}{2m}(xx'[0]+xx'[m])+\frac{1}{m}\sum_{p=1}^{m-1} xx'[p]$$

$$XX'[r] = \frac{1}{m}xx'[0]+\frac{2}{m}\sum_{p=1}^{m-1} xx'[p]\cos\left(\frac{\pi rp}{m}\right)+\frac{1}{m}xx'[m]\cos(r\pi)$$

$$0 < r < m$$

$$XX'[m] = \frac{1}{2m}(xx'[0]+(-1)^m xx[m])+\frac{1}{m}\sum_{p=1}^{m-1}(-1)^p xx[p] \ . \qquad (7.3.4)$$

Step 4 these line powers are now convolved with $H[r]$ to give $\hat{X}X[r]$. Note that the convolution in summation terms involves Δf. This may be combined with $H[r]$. Therefore, if the cosine bell is used for $h[p]$, $H_2[r] \times 1/2m\Delta t = 0\cdot25,\ 0\cdot50,\ 0\cdot25$, and the explicit relationships become, for the *one-sided* spectrum,

$$\hat{X}X[0] = 0\cdot5(XX'[0]+XX'[1])$$

$$\hat{X}X[r] = 0\cdot5XX'[r]+0\cdot25(XX'[r-1]+XX'[r+1])$$

$$\hat{X}X[m] = 0\cdot5(XX'[m-1]+XX'[m]) \ . \qquad (7.3.5)$$

These estimates are averages for bands $1/2m\Delta p$ wide, except at $r = 0$ and m where they are $1/4m\Delta p$ wide, centred at $f = r/2m\Delta p$, where r now is an integer varying between 0 and m. The diagram of $\hat{X}X[r]$ plotted against r resembles a histogram as shown in Figure 7.8.2.

7.4 Real relationship of the data transform to the variance spectrum

Equation (7.1.15) shows that an alternative method of calculating the variance spectrum is to transform the data directly. Again, however, Equation (7.1.15) cannot be applied as $x'[t]$ is not defined for $|t| > T/2$. Following the argument of Section 7.2 this problem is circumvented by multiplying $x'[t]$ by $h[t]$. Then Equation (7.1.15) reads

$$\hat{XX}[f] = \frac{1}{T}\left|\int_{-T/2}^{T/2} x'[t]h[t]\exp(-i2\pi ft)dt\right|^2 \tag{7.4.1}$$

$$= \frac{1}{n}|X'[f] * H[f]|^2 . \tag{7.4.2}$$

As $h[t]$ becomes very wide, i.e. n large, so $H[f]$ approaches a single spike and averaging over frequency reduces to a minimum:

$$X'[f] * H[f] = \hat{c}[f] = \hat{a}[f] - i\hat{b}[f] \tag{7.4.3}$$

from Equation (5.2.17). But $|\hat{c}[f]|^2 = \hat{c}[f]\hat{c}^*[f]$, therefore

$$|X'[f] * H[f]|^2 = \hat{a}^2[f] + \hat{b}^2[f] \tag{7.4.4}$$

and

$$\hat{XX}[f] = \frac{1}{n}\left(\hat{a}^2[f] + \hat{b}^2[f]\right) . \tag{7.4.5}$$

This is an even function and is multiplied by 2 if the one-sided transform is used. Also, if the variance density estimates are converted into average variance per frequency band (i.e. multiply by $1/n$), Equation (7.4.5) becomes

$$\hat{XX}[k] = \frac{1}{n} \times 2 \times \frac{1}{n}\left\{\left(\frac{n\hat{a}[k]}{2}\right)^2 + \left(\frac{n\hat{b}[k]}{2}\right)^2\right\}$$

$$= \frac{\hat{a}^2[k] + \hat{b}^2[k]}{2} , \tag{7.4.6}$$

which is equivalent to the equation for the variance of the periodic function, Equation (3.4.3), but the data have been modified by $h[t]$ and the estimate is an average for a band centred at k rather than a discrete estimate at k.

The bands are $1/n\Delta t$ wide except at $k = 0$ and $n/2$, where they are $1/2n\Delta t$ wide. Since $\hat{XX}[k]$, containing $n/2$ *full* bands, covers the same range of frequencies 0 to $1/2\Delta t$ as $\hat{XX}[r]$, which contains m full bands, it may be made equivalent to $\hat{XX}[r]$ by summing over $n/2m$ bands.

7.5 Practical steps using the direct transform

The major steps in calculating an estimate of the variance spectrum from a set of D observations of $x'[j]$ are as follows:

Step 1 remove mean and trends.

Step 2 apply the window function $h[j]$ now called tapering by Tukey (1967). He suggests multiplication of the first 10% by a rising cosine bell and the last 10% by a descending cosine bell. Let $G \approx D/10$, i.e.

$$
\begin{aligned}
h[j] &= \tfrac{1}{2}\left\{1 - \cos\left(\frac{\pi j}{G}\right)\right\} && 0 \leqslant j < G \\
&= 1 && G \leqslant j \leqslant D - G \\
&= \tfrac{1}{2}\left\{1 - \cos\left[\frac{\pi(D-j)}{G}\right]\right\} && D - G < j \leqslant D - 1 \ .
\end{aligned}
\tag{7.5.1}
$$

The 10% is arbitrary but should be between 5 and 25% (see Section 7.6.). Thus

$$
\hat{x}[j] = x'[j]h[j] \ .
\tag{7.5.2}
$$

Step 3 zeros may now be added to the end of $\hat{x}[j]$ to make factoring easy for the fast Fourier transform and to adjust the frequency bands to the desired widths and desired central frequencies (see fifth step). The number of zeros added is made equal to $n - D$ so that the total number of observations in $\hat{x}[j]$ is now n.

Step 4 calculate the Fourier coefficients $\hat{a}[f]$ and $\hat{b}[f]$ [Equations (5.2.12) and (5.2.13)] or, better still, the coefficients $\hat{a}[k]$ and $\hat{b}[k]$ [Equations (3.3.11) and (3.3.12)], which are related to $\hat{a}[f]$ and $\hat{b}[f]$ in Section 5, using the fast Fourier transform. Other methods may be used but normally they will be uneconomical.

Step 5 calculate the variances $\hat{XX}[k]$ and sum over blocks of 5 to 50, say $2z + 1$ centred at r, to give the $\hat{XX}[r]$ estimates, i.e.

$$
\begin{aligned}
\hat{XX}[0] &= \hat{a}^2[0] + \sum_{k=1}^{z} \frac{\hat{a}^2[k] + \hat{b}^2[k]}{2} \\
\hat{XX}[r] &= \sum_{k=r(2z+1)-z}^{r(2z+1)+z} \frac{\hat{a}^2[k] + \hat{b}^2[k]}{2} && 0 < r < m \\
\hat{XX}[m] &= \sum_{k=n/2-z}^{n/2-1} \frac{\hat{a}^2[k] + \hat{b}^2[k]}{2} + \hat{a}^2\left[\frac{n}{2}\right] \ .
\end{aligned}
\tag{7.5.3}
$$

In this particular example n has been chosen to be even, $2z + 1$ odd, and $(n/2)/m = 2z + 1$. Such a selection leads to frequency bands centred at

$$
0 \ , \ \frac{1}{2m\Delta t} \ , \ \frac{2}{2m\Delta t} \ , \ \cdots \ , \ \frac{r}{2m\Delta t} \ , \ \cdots \ , \ \frac{1}{2\Delta t} \ .
$$

However, there is no reason why other combinations should not be selected. In such cases the end frequency bands will be of various widths and care will be required in comparing them with the other bands. Like Equation (7.3.5), Equation (7.5.3) defines the *one-sided* spectrum.

7.6 Confidence intervals

The probability distribution of a parameter estimate may be approached in two ways. First, some theoretical distribution may be assumed on the basis of some *a priori* knowledge and, secondly, the distribution may be empirically determined by calculating several estimates from a series of experiments (samples). In the absence of a thorough knowledge of the process the best method is to take several samples as suggested, for example, by Schuster (1906b, p.72), Chapman and Bartels (1940, Section 16–26), Tukey (1949), and Jenkins (1961, p.163). Unfortunately the taking of several samples is usually impractical and there is no alternative to assuming some theoretical distribution which appears to apply. The simplest and most useful distribution is that of the Gaussian curve and this is the one which will be used here. Often the data are not normally distributed, so the use of this curve will be an approximation.

Since interest is centred on the variances rather than the data themselves, the actual distribution of interest will be the chi square distribution which will now be discussed.

7.6.1 The chi square distribution

The chi square distribution is given by

$$f(\chi^2) = \frac{(\chi^2)^{\nu/2-1}\exp(-\chi^2/2)}{2^{\nu/2}\Gamma(\nu/2)} \, , \tag{7.6.1.1}$$

where $\Gamma(n) = (n-1)!$ is the gamma function, and ν is known as degrees of freedom. Examples of χ^2 are given in Figure 7.6.1.1 for different ν.

This function is characterised by the following parameters:

mode $= \nu - 2$

mean $= \nu$

variance $= 2\nu$. $\tag{7.6.1.2}$

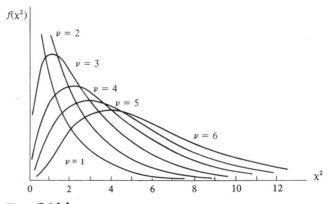

Figure 7.6.1.1.

If $a[0], a[1], ..., a[k], ..., a[n-1]$ are independent and normally distributed with a mean of $\bar{a}$, it is a well-known result (see Weatherburn, 1962, Chapter 9) that

$$\frac{n \sum_{k=0}^{n} (a[k]-\bar{a})^2/n}{\sigma^2} = \chi^2[n-1(\text{degrees of freedom})] \ , \qquad (7.6.1.3)$$

i.e.

$$\frac{n \times \text{sample variance}}{\text{population variance}} = \chi^2[n-1] \ . \qquad (7.6.1.4)$$

Usually there is some discussion as to n in the calculation of the sample variance and the number of degrees of freedom. If the sample variance is an unbiased estimate of σ^2, then it should be multiplied by $n-1$ in Equation (7.6.1.4) and then that equation becomes

$$\frac{\nu\hat{\sigma}^2}{\sigma^2} = \chi^2[\nu] \ . \qquad (7.6.1.5)$$

Equation (7.6.1.1) may be integrated and χ^2 has been tabled as a function of degrees of freedom and probability of being exceeded (cumulative chi square tables). Thus the limits of the left-hand side of Equation (7.6.1.5) may be specified if the number of degrees of freedom and the probability levels are given. For example, for $\nu = 10$ the left-hand side of Equation (7.6.1.5) will fall between $3 \cdot 94$ (= $\chi^2[10, 95\%]$) and $18 \cdot 31$ (= $\chi^2[10, 05\%]$) 90% of the time. If Equation (7.6.1.5) is rearranged, the limits of the population variance may be specified in terms of the sample variance, i.e.

$$\frac{\nu\hat{\sigma}^2}{\chi^2[\nu, \text{ smaller probability level}]} \leqslant \sigma^2 \leqslant \frac{\nu\hat{\sigma}^2}{\chi^2[\nu, \text{ larger probability level}]} \ . \qquad (7.6.1.6)$$

Because the multiplication factor, $\nu/\chi^2[\nu, \text{level}]$, is constant for $0 < r < m$, it is usually more convenient to plot the spectrum on a logarithmic scale. Then the confidence band is a constant width.

7.6.2 Application to the spectrum and degrees of freedom

In the calculation of the spectrum the sample variance in a given band is estimated by [Equation (7.5.3)]

$$\hat{\sigma}^2[r] = X\hat{X}[r] = \sum_{k=r(2z+1)-z}^{r(2z+1)+z} \frac{\hat{a}^2[k]+\hat{b}^2[k]}{2} \ . \qquad (7.6.2.1)$$

If $\hat{a}[k]$ and $\hat{b}[k]$, which are derived from an orthogonal transformation, are assumed to be normally distributed, they are independent and the confidence bands of $\hat{\sigma}^2[r]$ of Equation (7.6.2.1) may be calculated with the use of Equation (7.6.1.6). According to Tukey (1967), tapering reduces the total degrees of freedom by G, the number of points involved at either end. Then, if the $n/2$ elementary bands, $(\hat{a}^2[k] + \hat{b}[k])/2$, are summed into non-overlapping groups of $2z + 1$, the number of degrees of freedom for each variance estimate will be

$$\nu \approx \frac{(2z + 1)(D - G)}{n/2} , \qquad (7.6.2.2)$$

except for $r = 0,n$ where only $z + 1$ elementary bands are summed.

When the spectrum is calculated via the autocovariances, the degrees of freedom must be approached differently since the estimates $\hat{\sigma}^2[r]$ are not independent because of blurring created by the spectral windows (see Blackman and Tukey, 1958, Sections 6 and 7; Jenkins and Watts, 1968, Section 6.3). Following Tukey (1949) the degrees of freedom are calculated using the parameters given in Equation (7.6.1.2), which give

$$\text{coefficient of variation or r.m.s.} = (2/\nu)^{\frac{1}{2}} \qquad (7.6.2.3)$$

or

$$\frac{(\text{average})^2}{\text{variance}} = \frac{\nu}{2} .$$

As was shown with Equation (7.2.7)

$$\text{ave } \hat{XX}[f_1] = \int_{-\infty}^{\infty} XX[f]H[f_1 - f]\,\mathrm{d}f .$$

This states that the ensemble average of the estimate is equal to the weighted integral over frequency, i.e. the sum of the filtered true spectrum at discrete frequencies. If the filtered true spectrum $\tilde{XX}[f] = XX[f]H[f_1 - f]$, then

$$\text{ave } \hat{XX}[k_1] = \sum \tilde{XX}[k]\Delta f . \qquad (7.6.2.4)$$

Similarly it may be shown that the ensemble variance is approximately equal to the integral with respect to frequency of the squared filtered spectrum divided by $n\Delta t$, i.e.

$$\text{var } \hat{XX}[f_1] = \frac{1}{n\Delta t}\int_{-\infty}^{\infty} \tilde{XX}^2[f]\,\mathrm{d}f \qquad (7.6.2.5)$$

or

$$\text{var } \hat{XX}[k_1] \doteq \frac{1}{n\Delta t}\sum \tilde{XX}^2[k]\Delta f . \qquad (7.6.2.6)$$

Thus Equation (7.6.2.3) becomes

$$\nu = \frac{2n\Delta t\left\{\int_{-\infty}^{\infty} \tilde{XX}[f]\,\mathrm{d}f\right\}^2}{\int_{-\infty}^{\infty} \tilde{XX}^2[f]\,\mathrm{d}f} \qquad (7.6.2.7)$$

where $\left\{ \int_{-\infty}^{\infty} X\tilde{X}[f] \, df \right\}^2 \bigg/ \int_{-\infty}^{\infty} X\tilde{X}^2[f] \, df$ is known as the equivalent width of $X\tilde{X}[f]$, a function of the spectral window. Jenkins (1961), for example, shows that Equation (7.6.2.7) may be written

$$\nu = \frac{2T}{\int_{-\infty}^{\infty} h[t] \, dt} \qquad\qquad\qquad (7.6.2.8)$$

and has listed ν, called equivalent degrees of freedom, for various windows $h[t]$.

For the discrete spectrum Equation (7.6.2.7) becomes

$$\nu = \frac{2n\Delta t \left(\sum X\tilde{X}[k] \right)^2 \Delta f^2}{\sum X\tilde{X}^2[k] \, \Delta f} \qquad\qquad (7.6.2.9)$$

but $\Delta f = 1/n\Delta t$, so

$$\nu = \frac{2 \left(\sum X\tilde{X}[k] \right)^2}{\sum X\tilde{X}^2[k]} . \qquad\qquad (7.6.2.10)$$

To account for both a's and b's Tukey makes an elementary bandwidth equal to $\Delta f = 1/2n\Delta t$. Consequently the 2 is missing from his equation which is equivalent to Equation (7.6.2.8) (Blackman and Tukey, 1958, p.24).

For the hanning window and the autocovariance approach $\nu \approx 2n/m$. Since the maximum total degrees of freedom are n and there are $m+1$ estimates, these estimates are not independent. In fact, only alternative estimates are independent, giving $m/2$ equivalent independent estimates in all for the hanning window.

Equation (7.6.2.9) shows that, for an ideal spectral window (one which is uniform over the elementary bands in the equivalent width and zero outside) and for a true spectrum which is flat, the number of degrees of freedom is equal to the number of elementary bands. However, when the elementary bands are unequal, then ν is proportionately reduced {put $X\tilde{X}[k]$ of different magnitudes in Equation (7.6.2.10)}. In the limiting case of a single significant elementary band $\nu = 2$ (two estimates $+f$ and $-f$, or amplitude and phase, or $\hat{a}[k]$ and $\hat{b}[k]$). Of course, the end estimates $X\hat{X}[0]$ and $X\hat{X}[m]$ have only half as many degrees of freedom as the other estimates.

The last equation shows that reducing the number of final estimates increases the number of degrees of freedom so long as the variances in the elementary bands are approximately equal. Therefore there is a trade off between the width of the confidence bands and the resolution. On the other hand, if one peak stands out, adding more elementary bands will not change ν. An alternative approach is to calculate two extreme spectra using inner and outer windows (Wonnacott, 1961; Tukey, 1967).

Note that these results are only approximate. Yet only approximate results are needed according to Blackman and Tukey (1958, p.21), since usually all the assumptions will not be fulfilled. Nevertheless this discussion has revealed some of the relationships between the variability of the true spectrum, the windows, and the confidence, bias, and resolution of the estimates.

In the preparation for the calculation of a spectrum the interrelated variables D, n, m, Δt, ν, $r/2m\Delta t$, $2z+1$, and $h[j]$ must be selected. Usually some will be already fixed, some will be prescribed by the nature of the problem, and others will be unspecified. In each problem the needs, for example in terms of width of confidence bands, must be modified to an acceptable point where they will fit together under the restraints of the fixed elements and interrelating equations. Certain rules of thumb may be given such as $D/20 \leqslant n - D \leqslant D/4$, but each problem must be solved independently. Blackman and Tukey (1958) and Jenkins and Watts (1968) give a number of practical examples.

7.7 Prewhitening and frequency rejection

From the discussion in Section 7.6.2 it is clear that the maximum number of degrees of freedom are available when the spectrum is relatively flat. Also, it should be clear that the effect of side lobes of the spectral windows is a minimum when the spectrum is relatively uniform. Consequently an attempt should be made to modify the data so that the spectrum is flat. Once the transform and window(s) have been applied with minimum blurring, inverse modification can replace the unevenness in the spectrum. This suggests that the spectrum should be calculated initially from the raw data so that gross features of the spectrum are revealed. Then the data should be modified so as to remove these features, and the spectrum recalculated followed by inverse modification.

As a relatively flat spectrum is that produced by white noise, the initial modification is known as prewhitening. The inverse is post-darkening. Most spectra decline towards high frequencies, so the prewhitening routine usually requires the reduction of low frequencies and amplification of high frequencies. This may be successfully accomplished by the following procedure:

$$x[j_1] = x[j_1] - Ex[j_1 - 1] = \sum w[j] x[j_1 + j] \ , \tag{7.7.1}$$

where the weighting function

$$w[j] = 1 \qquad \text{for } j = 0$$
$$\quad = -E \qquad (E < 1) \text{ for } j = -1$$
$$\quad = 0 \qquad \text{for } -1 > j > 0. \tag{7.7.2}$$

The response of this weighting function is

$$W^*[f] = 1 \exp(i2\pi f 0) - E \exp\{i2\pi f(-1)\} = 1 - E \exp(-i2\pi f) \tag{7.7.3}$$

and its modulus is

$$|W[f]| = [\{1 - E \exp(i2\pi f)\}\{1 - E \exp(-i2\pi f)\}]^{\frac{1}{2}}$$
$$\quad = \{(1 + E^2) - 2E \cos(2\pi f)\}^{\frac{1}{2}}. \tag{7.7.4}$$

It may be seen that as $E \rightarrow 1$ the effect of this function is increased. From the other point of view, the spectra which have a large (small) ratio of the variance in the first frequency to that in the last require large (small) E to convert them to a flat spectrum.

Post-darkening the a's and b's requires division by Equation (7.7.3), a complex variable. For example, if $\dot{a}[k]$ and $\dot{b}[k]$ are the prewhitened a's and b's, the post-darkened b is given by

$$\hat{b}[k] = \frac{E \sin(2\pi k/n)\dot{a}[k] + \{1 - E \cos(2\pi k/n)\}\dot{b}[k]}{\{1 - E \cos(2\pi k/n)\}^2 + \{E \sin(2\pi k/n)\}^2}. \tag{7.7.5}$$

If the variance spectrum is post-darkened, then division must be by the square of Equation (7.7.4) which is real.

Other procedures may be required to flatten the spectrum. Where sharp peaks occur, they should be removed by a band-pass filter. It should be noted that the practice of calculating first differences is the limiting form of prewhitening with $E = 1 \cdot 0$.

7.8 An example of spectral analysis

Table 7.8.1 lists the precipitation totals in inches for State College, Pennsylvania, for 80 consecutive days from 6 May 1960. For comparative purposes a harmonic analysis was performed and the results presented in Figure 7.8.1. This shows the percentage variance contributed by each of the 40 calculable frequencies. Their distribution suggests that frequency 6 and its harmonics are the most significant. In fact, periods between 11 and 16 days account for over 21% of the variance and the conclusion might be drawn that there was possibly a generating process at the frequency 6. This would have the effect of channelling further research into the study of that particular scale. However, it is very difficult to support the assumption that rainfall is periodic particularly at a basic interval of 80 days. The obvious alternative assumption is one of non-periodicity and the application of the steps given in Section 7.5. With

Table 7.8.1. Precipitation totals in inches for State College, Pa., for 80 consecutive days from 6 May 1960. Read by column.

0·00	0·50	0·00	0·00	0·00
0·00	1·23	0·00	0·00	0·00
0·30	0·05	0·00	0·10	0·28
1·66	0·00	0·00	0·00	0·02
0·32	0·00	0·00	0·00	0·00
0·57	0·00	0·01	0·00	0·55
0·00	0·00	0·44	0·22	1·03
0·25	0·78	0·18	0·00	0·00
0·16	0·05	0·50	0·00	0·00
0·05	0·58	0·07	0·76	0·00
0·05	0·00	0·00	0·00	0·17
0·00	0·00	0·37	0·28	0·09
0·45	0·15	0·00	0·00	0·00
0·00	0·00	0·05	0·00	0·00
0·06	0·00	0·00	0·00	0·00
0·18	0·06	0·05	0·00	0·03

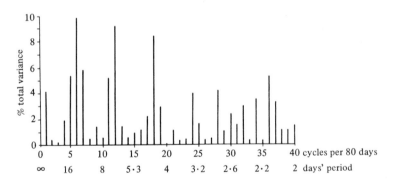

Figure 7.8.1.

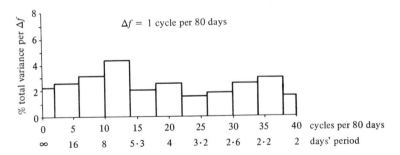

Figure 7.8.2.

$D = 80$, 10% tapering, the addition of 20 zeros, and summation over 5 bands the result is given in Figure 7.8.2.

Immediately it becomes clear that no frequencies stand out as they did in the periodic spectrum the confidence bands (Exercise 9.9.4) confirm that none is statistically significant at the 90% level. On the other hand, the band centred at 12 cycles per 80 days, containing $16 \cdot 5\%$ of the variance, appears to be more important than the others. Such a result could not have been foreseen from the periodic analysis. As is usually the case the reality of a peak cannot be confirmed from one set of data. More samples must be collected or other forms of analysis must be performed. A peak at about seven days' period in these data is physically reasonable, since this is a possible frequency of cyclones when the ratio of their speed of movement to their spacing remains relatively constant [Equation (1.2.3)]. Yet the peak should not be expected from all sample periods as the ratio does not remain constant over long time intervals. Also, other generating processes may be operating. Consequently, in problems such as the one under consideration, interest may be centred upon the non-stationarity of the series and the specific processes acting during a particular observation period. In which case the collection and spectral analysis of further samples from other periods may be irrelevant.

One step which should not be omitted is a consideration of the phases of the elementary bands. Non-random generating processes will tend to produce uniformly varying phases in groups. Here the result of the phase breakdown is negative as the five elementary bands centred at 12 cycles per 80 days show widely fluctuating phases. They are 55, 149, 194, 316, and 273 degrees (see Exercise 7.9.3).

From the results in Figure 7.8.2 it may be concluded only that (1) a wide range of frequencies is present, and that (2) there is a suggestion of a peak at *about* seven days' period which, from further analysis, might prove to be physically significant. What the further analysis might be will depend upon the nature of the associated data and the various hypotheses suggested. In this example, if the generating process is believed to be related to the pressure patterns, a correlation between rainfall and pressure as a function of frequency (cross spectral analysis) might be performed.

7.9 Exercises

7.9.1 Calculate the periodic percentage variance spectrum of every alternate observation listed in Table 7.8.1. Compare with Figure 7.8.1 to see the effect of aliasing.

7.9.2 Plot the modulus of the prewhitening function for twelve points for (a) $E = 0 \cdot 6$ and (b) $E = 0 \cdot 9$.

7.9.3 From the phases, given below of the first twenty-three elementary bands in the data of Table 7.8.1, plot separate harmonic dials (Chapman and Bartels, 1940, Section 16.18) for each of the corresponding bands shown in Figure 7.8.2.

	Phase		Phase		Phase		Phase
0	–	6	96	12	337	18	61
1	4	7	336	13	55	19	97
2	113	8	130	14	149	20	225
3	103	9	114	15	194	21	222
4	268	10	254	16	316	22	200
5	357	11	212	17	273		

7.9.4 Calculate the confidence bands for Figure 7.8.2 and plot on a logarithmic scale. The percentage variance estimates are $2 \cdot 2$, $2 \cdot 5$, $3 \cdot 1$, $4 \cdot 1$, $2 \cdot 0$, $2 \cdot 6$, $1 \cdot 5$, $1 \cdot 8$, $2 \cdot 5$, $2 \cdot 9$, $1 \cdot 6$.

7.9.5 Reading: Tukey (1949, 1961, 1967); Jenkins (1961). Refer to: Bendat and Piersol (1966); Blackman and Tukey (1958); Granger and Hatanaka (1964); Jenkins and Watts (1968).

8

Cross spectral analysis

8.1 Introduction

So far consideration has been given to the one variable case and discussion has moved from periodic to non-periodic data and from looking at amplitudes to looking at variances.

In fact, the breakdown of the variance into its frequency components is significant on two counts.

1 It is basic to the description of contiguous and dependent data. The spectrum is better than the autocovariances as it is more clearly understood and more easily judged for its statistical reliability. The mean and spectrum, then, are the principle parameters for describing sequenced data. As in simple statistical description the calculation of higher moments may be necessary, thus rendering polyspectra, but these are omitted from the present discussion. It should be noted here that components of variance calculated for the statistical description need not have a meaning in the real world.

2 It is also an important method of analysing the variance as a step towards isolating scale phenomena in the real physical and social world. In other words, certain variance components might be associated with particular physical or social mechanisms.

In order to take this idea further why not calculate the degree of association between phenomena? In its simplest form for independent (unsequenced) data this is done by linear regression:

$$y = u + vx .$$

$$(8.1.1)$$

With sequenced data, y, u, v, and x are functions of t, space, or time. Consequently, y will most likely be a function of all the x's in the region adjacent to y. For a time series y will be associated with past and present x only and Equation (8.1.1) may be rewritten

$$y[t] = u[t] + \int_0^\infty v[q]x[t-q]\,dq .$$

$$(8.1.2)$$

Alternatively, if the sequence has no direction such as in the case of most spatial data, then $y[t]$ will be influenced by $x[t]$ on either side and Equation (8.1.2) becomes

$$y[t] = u[t] + \int_{-\infty}^\infty v[q]x[t+q]\,dq .$$

$$(8.1.3)$$

8.2 Solution of the regression equation

This equation, (8.1.3), may be solved as a function of frequency. If $x[t]$ and $u[t]$ are uncorrelated, both sides of Equation (8.1.2), when multiplied by $x[t+p]$ and averaged over all t, give

$$\lim_{T\to\infty}\frac{1}{T}\int_{-\infty}^{\infty}y[t]x[t+p]\,dt = \int_{-\infty}^{\infty}v[q]\lim_{T\to\infty}\frac{1}{T}\int_{-\infty}^{\infty}x[t+q]x[t+p]\,dt\,dq$$

(8.2.1)

or

$$xy[p] = \int_{-\infty}^{\infty}v[q]xx[p-q]\,dq \ .$$

(8.2.2)

Transform, then,

$$XY[f] = V[f]XX[f]$$

(8.2.3)

and

$$V[f] = \frac{XY[f]}{XX[f]} \ ,$$

(8.2.4)

which is the response function of the system and is analogous to v in the ordinary regression equation, i.e.

$$v = \frac{\sum xy/n}{\sum x^2/n} \ ,$$

(8.2.5)

but now Equation (8.2.4) is complex and is a function of frequency.

Similarly a coefficient analogous to r_{yx}^2, the coefficient of determination or the square of the correlation coefficient,

$$r_{yx}^2 = \frac{\left(\sum xy/n\right)^2}{\left(\sum x^2/n\right)\left(\sum y^2/n\right)} \ ,$$

(8.2.6)

may be defined as

$$R_{yx}^2[f] = \frac{XY^2[f]}{XX[f]YY[f]} \ ,$$

(8.2.7)

which is known as the coherence or coherency and has a range between 0 and 1. It should be noted that the terms coherence and coherency have also been applied to the square root of Equation (8.2.7).

The equations for $XX[f]$ and $YY[f]$ have already been given in Chapter 7 and by analogy the equations for $XY[f]$ may be developed in the same way, Equations (7.1.10)-(7.1.14). Thus

$$xy[p] = \lim_{T\to\infty}\frac{1}{T}\int_{-\infty}^{\infty}y[t]x[t+p]\,dt$$

(8.2.8)

may be transformed to give

$$XY[f] = \int_{-\infty}^{\infty} \left(\lim_{T \to \infty} \frac{1}{T} \int_{-\infty}^{\infty} y[t] x[t+p] \, dt \right) \exp(-i2\pi fp) \, dp \qquad (8.2.9)$$

$$= \lim_{T \to \infty} \frac{1}{T} Y[f] X^*[f] \qquad (8.2.10)$$

$$= \lim_{T \to \infty} \frac{1}{T} \left\{ \int_{-\infty}^{\infty} y[t] \exp(-i2\pi ft) \, dt \right\} \left\{ \int_{-\infty}^{\infty} x[t] \exp(i2\pi ft) \, dt \right\} . \qquad (8.2.11)$$

8.3 Explicit equations for cross spectral analysis (covariance approach)

$XX[f]$ is the variance spectrum as calculated in Chapter 7 and therefore must be replaced by the estimate $\hat{X}\hat{X}[f]$ or $\hat{X}\hat{X}[r]$.

It will be assumed here that both series have a zero mean and are trend free, and that any initial smoothing and prewhitening will have been performed. Although it is not necessary, similar smoothing and prewhitening functions should be used on each series. This simplifies the inverse smoothing and post-darkening.

From Equation (8.2.1) it can be seen that $xy[p]$ are the lag covariances defined in the same way as $xx[p]$ except that the summation must be calculated over negative lags because the function is not even, i.e.

$$xy'[p] = \sum_{-(n-|p|)}^{n-|p|} \frac{y[t] x[t+p]}{n - |p|} . \qquad (8.3.1)$$

There has been some discussion in the literature, concerning the denominator in this equation and also in Equation (7.2.1) which suggests that $n - |p|$ should be replaced by n. This substitution is easily made and should not affect the understanding of the following.

The transformation of $xy'[p]$ in $XY'[f]$ will require both cosines and sines so the $xy'[p]$ is broken into even and odd series according to Equations (4.8.1.1) and (4.8.1.2):

$$xy'_E[p] = \tfrac{1}{2}(xy'[p] + xy'[-p]) , \qquad (8.3.2)$$

$$xy'_O[p] = \tfrac{1}{2}(xy'[p] - xy'[-p]) . \qquad (8.3.3)$$

The transformation of $xy'_E[p]$ is performed with the same equations as were used on $xx'[p]$, i.e. Equations (7.3.4) and (7.3.5) followed by any inverse smoothing and post-darkening. The resulting estimates $\hat{X}\hat{Y}_E[r]$ are known collectively as the cospectrum of $y[t]$ and $x[t]$. They give the frequency breakdown of the instantaneous covariance between $y[t]$ and $x[t]$. As a distinct function the cospectrum has direct physical application in areas such as meteorology where, if one of the variables is wind, the covariance is proportional to the transport of the other variable (see Rayner, 1967b). For more general application the cospectrum is considered as just an intermediate function required for the estimation of regression statistics analogous to slope and correlation.

For the $xy'_O[p]$ line powers are available for $0 < r < m$ and are given by

$$XY'_O[r] = \frac{2}{m} \sum_{p=1}^{m-1} xy'_O[p] \sin\left(\frac{pr\pi}{m}\right), \qquad 0 < r < m . \tag{8.3.4}$$

Equation (7.3.5) plus inverse smoothing and post-darkening follows as for $\hat{XY}_E[r]$. This function is known as the quadrature spectrum. It is the spectrum of the 'out-of-phase' portion of the covariance between $y[t]$ and $x[t]$. Alone it is seldom used.

The $\hat{XY}_E[r]$ and $\hat{XY}_O[r]$ are analogous to the $\hat{a}[k]$ and $\hat{b}[k]$ of harmonic analysis. Consequently the cross covariance is given by

$$|\hat{XY}[r]| = (\hat{XY}_E^2[r] + \hat{XY}_O^2[r])^{\frac{1}{2}} \tag{8.3.5}$$

and the phase $\Phi_{yx}[r]$, the average distance between the maximum of $x[t]$ at frequency r and the maximum of $y[t]$ in the same frequency band, is given by

$$\hat{\Phi}_{yx}[r] = \arctan\left(\frac{\hat{XY}_O[r]}{\hat{XY}_E[r]}\right) \tag{8.3.6}$$

where y follows x in time or y is further from the origin than x. It may be put in terms of the units of Δt, known as lag τ_{yx}, if it is remembered that angular distance from the origin is phase shift, $\Phi_{yx}[r]/r$, and that there are $2m\Delta t$ units in 2π angular units, i.e.

$$\tau_{yx}[r] = \frac{\Phi_{yx}[r]}{r} \frac{2m\Delta t}{2\pi} , \tag{8.3.7}$$

which is equivalent to Equation (3.2.7). The response function of Equation (8.2.4) therefore contains two parts, the magnitude, known as the gain, which is more closely related to v [Equation (8.2.5)], the proportionate increase in y for an increase in x,

$$\hat{V}_{yx}[r] = \frac{|\hat{XY}[r]|}{\hat{XX}[r]} , \tag{8.3.8}$$

and the phase or lag given by Equation (8.3.6) or (8.3.7). Just as in simple linear regression there is also the regression of x on y which produces $\hat{V}_{xy}[r]$:

$$\hat{V}_{xy}[r] = \frac{|\hat{XY}[r]|}{\hat{YY}[r]} . \tag{8.3.9}$$

Similarly the coherence may be written

$$\hat{R}_{yx}^2[r] = \frac{\hat{XY}_E^2[r] + \hat{XY}_O^2}{\hat{XX}[r]\,\hat{YY}[r]} , \tag{8.3.10}$$

which is the same as the product of the gains (regression coefficients).

8.4 Explicit equations for cross spectral analysis (fast Fourier transform approach)

It turns out that the various parameters discussed above are very easily obtained from the amplitude estimates of the elementary bands of each separate series $x[t]$ and $y[t]$. This reduces the number of calculations considerably since no lag covariances are required. This is particularly important where several cross spectra are required each involving the same $x[t]$. The estimates for $x[t]$ need be calculated only once.

From Equation (8.2.11) it is clear that $XY[f]$ is the product of the individual transforms of $x[t]$ and $y[t]$. If each is multiplied by the tapering function (equivalent of the lag window), Equation (8.2.11) becomes, following the same argument as in Section 7.4,

$$\hat{XY}[f] = \frac{1}{T} \hat{YY}[f] \hat{XX}^*[f] \tag{8.4.1}$$

and for a one-sided function

$$\hat{XY}[k] = \tfrac{1}{2}(\hat{a}_y[k] - i\hat{b}_y[k])(\hat{a}_x[k] + i\hat{b}_x[k]), \tag{8.4.2}$$

which may be rearranged into real and imaginary parts

$$= \frac{\hat{a}_x[k]\hat{a}_y[k] + \hat{b}_x[k]\hat{b}_y[k]}{2} - i\frac{(\hat{a}_x[k]\hat{b}_y[k] - \hat{a}_y[k]\hat{b}_x[k])}{2}, \tag{8.4.3}$$

where each a and b is obtained according to steps 1 to 4 of Section 7.5. Therefore the cospectrum

$$\hat{XY}_E[k] = \frac{\hat{a}_x[k]\hat{a}_y[k] + \hat{b}_x[k]\hat{b}_y[k]}{2} \tag{8.4.4}$$

and the quadrature spectrum

$$\hat{XY}_O[k] = \frac{\hat{a}_x[k]\hat{b}_y[k] - \hat{a}_y[k]\hat{b}_x[k]}{2}. \tag{8.4.5}$$

Next these are summed over $2z + 1$ elementary bands to produce the final estimates, i.e.

$$\hat{XY}_E[r] = \sum_{k=r(2z+1)-z}^{r(2z+1)+z} \frac{\hat{a}_x[k]\hat{a}_y[k] + \hat{b}_x[k]\hat{b}_y[k]}{2} \tag{8.4.6}$$

and

$$\hat{XY}_O[r] = \sum_{k=r(2z+1)-z}^{r(2z+1)+z} \frac{\hat{a}_x[k]\hat{b}_y[k] - \hat{a}_y[k]\hat{b}_x[k]}{2}. \tag{8.4.7}$$

As noted previously in Section 7.5, step 5, it is not necessary that the summation be over an odd number $(2z + 1)$ of elementary bands. An odd number is convenient so long as it is a factor of $n/2$, since the final end bands are then exactly half the width of the other final bands. Again, with n even, $\hat{\bar{a}}^2[0]$ and $\hat{\bar{a}}^2[n/2]$ will have double the weight of the other

elementary frequencies and end band equations may be written for both
Equations (8.4.6) and (8.4.7) corresponding to those in Equation (7.5.3).
The results of Equations (8.4.6) and (8.4.7) may now be used as input
to Equations (8.3.5) to (8.3.10) for the estimates of cross covariance,
phase, lag, gain, and coherence.

8.5 Confidence bands for the cross spectra

It was pointed out in Section 7.6 that confidence bands for the variance
spectrum are only approximate. The same holds true for the following
statistical limits of the coherence, phase, and gain. For a full discussion
of the relationships reference should be made to Jenkins and Watts
(1968). The degrees of freedom ν will be the same as those for the
variance spectrum and will be dependent upon the window. Again bias
and confidence will be inversely related.

8.5.1 Coherence

It may be shown that $\text{arctanh}(|\hat{R}_{yx}[r]|)$, where

$$\text{arctanh}(|\hat{R}_{yx}[r]|) = \tfrac{1}{2}\ln\left(\frac{1+|\hat{R}_{yx}[r]|}{1-|\hat{R}_{yx}[r]|}\right) , \qquad (8.5.1.1)$$

is normally distributed with a variance approximately equal to $1/\nu$, i.e.

$$\text{var}(\text{arctanh}(|\hat{R}_{yx}[r]|)) \approx \frac{1}{\nu} . \qquad (8.5.1.2)$$

Therefore, if $\pm g[\%]$ are the expected limits, for a given probability level,
of a normally distributed variable (e.g. $g[80\%] = 1\cdot282, g[90\%] = 1\cdot645,$
$g[95\%] = 1\cdot960$) the limits of Equation (8.5.1.1) may be calculated.
That is, the approximate limits will be

$$\text{arctanh}(|\hat{R}_{yx}[r]|) \pm g[\%]\left(\frac{1}{\nu}\right)^{\frac{1}{2}} . \qquad (8.5.1.3)$$

As these limits are constant on an arctanh scale it is more practical to plot
the transformed coherence [Equation (8.5.1.1)] than to retransform each
pair of limits [Equation (8.5.1.3)]. When a small number of degrees of
freedom is available, it should be noted (Tukey, 1967) that a completely
incoherent pair of series will show an average coherence of $2/\nu$.

8.5.2 Phase

The ratio $\hat{XY_O}[r]/\hat{XY_E}[r]$ also approximates a normal distribution with

$$\text{var}(\tan\Phi_{yx}[r]) \approx \sec^4\Phi_{yx}[r]\frac{1}{\nu}\left(\frac{1}{R_{yx}^2[r]}-1\right) , \qquad (8.5.2.1)$$

where sec refers to secant $= 1/\text{cosine}$. Therefore, following the argument
in Section 8.5.1 the limits are given by

$$\tan\Phi_{yx}[r] \pm g[\%]\left\{\sec^4\Phi_{yx}[r]\frac{1}{\nu}\left(\frac{1}{R_{yx}^2[r]}-1\right)\right\}^{\frac{1}{2}} , \qquad (8.5.2.2)$$

where $R_{yx}^2[r]$ must be replaced by $\hat{R}_{yx}^2[r]$. Jenkins and Watts (1968, Figure 9.3) have plotted a simple graph of these limits as a function of $R_{yx}^2[r]$ and ν.

More precise statistical limits may be obtained from

$$\Phi_{yx}[r] \pm \arcsin\left\{\frac{2}{\nu-2}F[2,\nu-2;100-\%]\left(\frac{1-\hat{R}_{yx}^2[r]}{\hat{R}_{yx}^2[r]}\right)\right\}^{\frac{1}{2}}, \qquad (8.5.2.3)$$

where $F[2,\nu-2;100-\%]$ refers to the F distribution with two different degrees of freedom and a given confidence level. For example, with $\nu = 14$ and $\% = 90\%$, $F[2,12;10\%] = 2\cdot8068$ and, with $\nu = 22$ and $\% = 95\%$, $F = 3\cdot4928$. It should be noted that the confidence bands for phase (and for gain) are dependent upon the uncontrollable coherence as well as degrees of freedom.

8.5.3 Gain

Similar in form to the phase confidence limits, those for gain are

$$\hat{V}_{yx}[r] \pm \hat{V}_{yx}[r]\left\{\frac{2}{\nu-2}F[2,\nu-2;100-\%]\left(\frac{1-\hat{R}_{yx}^2[r]}{\hat{R}_{yx}^2[r]}\right)\right\}^{\frac{1}{2}}. \qquad (8.5.3.1)$$

8.6 An example of cross spectral analysis

In this example the data are a year (13 August 1962–12 August 1963) of surface atmospheric pressure observations at two stations in the Southeastern Pacific. Chatham Island at 43° 58′ S, 176° 33′ W lies some 550 miles to the east and downstream, in terms of atmospheric flow, from Christchurch at 43° 29′ S, 172° 32′ E. The original 6-hourly observations were converted to daily observations by averaging, thereby reducing aliasing and the amount of computation, which was performed by the lag product technique (Rayner, 1965).

The variance and coherence spectrum with 90% confidence bands are presented alongside the lag spectrum in Figure 8.6.1. Both variance spectra show relatively strong peaks in the frequencies centred at 5 and 10 cycles per 80 days (16 and 8 days' period), which suggests the possibility of similar generating processes. Indeed, this is supported by the coherence spectrum which gives $|\hat{R}_{yx}[r]|$ values of $0\cdot88$ and $0\cdot84$. Physically these peaks and correlations may be explained by the occurrence of major atmospheric pressure disturbances which are observed to move from west to east in association with midlatitude westerly circulation. If this is so, then a disturbance at Christchurch must be observed at some later time at Chatham Island. The average interval of time which elapses between similar events in a particular frequency band is given by the phase or lag spectrum. As will be evident, the lag spectrum shows a high degree of consistency even in frequencies where coherence is low. The average lag time is about one day and at frequencies 5 and 10 is estimated to be $1\cdot0$ and $1\cdot2$ days respectively. The 90% confidence bands are in the order of

±0·5 days. These results indicate that pressure disturbances travel on the average at about 550 miles per day in this region. It might be argued that such an outcome was to be expected from previous individual case studies. However, the result of this study is more general since it has taken into account all disturbances of all calculable scales in that one year period. Case studies refer only to specific scales and short time intervals. Furthermore, they may not be typical. If the gain spectrum were

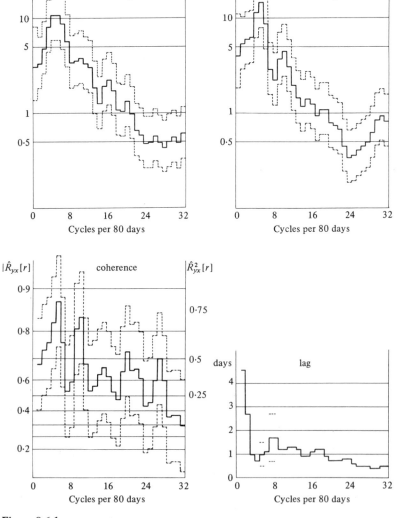

Figure 8.6.1.

included, it would show the relative intensification or decay of the pressure disturbances at each scale as they moved from one station to the other. For example, at 16 and 18 days the cross covariances are $10 \cdot 9$ mb^2 and $3 \cdot 48$ mb^2 and the variances at Christchurch are $0 \cdot 73$ mb^2 and $4 \cdot 49$ mb^2, which give gains of $1 \cdot 01$ and $0 \cdot 77$ respectively. Therefore pressure variations with approximately 16-day periods show little change in their passage from Christchurch to Chatham Island whereas those at 8-day periods tend on the average to decay.

A few comments are pertinent concerning the two cross spectra presented. With 18 degrees of freedom the coherence spectrum, which is plotted on an arctanh scale, presents wide confidence bands. In turn the coherence estimate appears in the calculations for the confidence bands of the phase spectrum. Hence the latter confidence bands are less reliable and have been calculated for only two frequencies in Figure 8.6.1. In converting phase estimates to lag estimates it should be remembered that although the phase is given for only the first cycle, it may refer to the second, third, etc., cycles. For instance, for the period of 8 days the lag has been accepted as $1 \cdot 2$ days, yet in fact there is nothing in the data sequences to prevent this from being $9 \cdot 2$, $17 \cdot 2$, etc., days. Other factors must be considered before the first cycle is recognised as the significant one.

Other examples of cross spectra are given in Rayner (1967b) and in the references (Appendix B). Cross spectral analysis may be employed in any problem where regression is used and where the data are sequenced. For the application of the technique to multivariate processes in general reference should be made to Jenkins and Watts (1968).

8.7 Exercises

8.7.1 If two sequences may be represented by cosine curves of the same frequency, what is the simple correlation coefficient between them? Hint: let $x[j] = A_x[k] \cos\{(2\pi jk/n) - \Phi_x[k]\}$
and $y[j] = A_y[k] \cos\{(2\pi jk/n) - \Phi_y[k]\}$.

8.7.2 What would be the simple correlation coefficient of the example in Exercise 8.7.1 if $\Phi_y[k] - \Phi_x[k] = \pi/2$? What would be the coherence in the same situation? Hint: let $a_x[k]$, $b_x[k]$, $a_y[k]$, and $b_y[k]$ be the components of $A_x[k]$ and $A_y[k]$.

8.7.3 Calculate the two sets of 90% confidence bands [Equations (8.5.2.2) and (8.5.2.3)] for the following estimated phase angles and coherences for 17 degrees of freedom; (a) $\Phi_{yx} = 1°$, $\hat{R}_{yx}^2 = 0 \cdot 9$; (b) $\Phi_{yx} = 1°$, $\hat{R}_{yx}^2 = 0 \cdot 6$; (c) $\Phi_{yx} = 89°$, $\hat{R}_{yx}^2 = 0 \cdot 9$; (d) $\Phi_{yx} = 89°$, $\hat{R}_{yx}^2 = 0 \cdot 6$.

8.7.4 Obtain two separate series of random numbers and compare their spectra and cross spectra.

8.7.5 Reading: Jenkins (1962, 1963, 1965); Jenkins and Watts (1968); Jones (1965); Tukey (1967).

Two-dimensional analysis

9.1 Introduction

By 'two dimensional' is meant that the dependent variable, any x, is a function of two coordinates in a Cartesian system. Up to this point the sequenced variable is assumed to be a function of one coordinate, time or distance, which requires one subscript. Now, as a function of two-dimensional space, or space–time, it requires two identification labels t_1 and t_2 (or j_1 and j_2). Examples of such variables are the elevation of land above sea level on a flat earth, population densities on a flat earth, the density of emulsion on a photograph, and the time variation of temperature along a traverse.

It is unimportant which coordinate is labelled t_1 and which is labelled t_2 so long as the results are viewed in the same system as the equations remain unchanged. In the following the written page format will be used with t_1 increasing to the right and t_2 increasing down the page.

As in the one-dimensional case either periodicity or a series of zeros may be assumed beyond the limits of observation. From the preceding chapters it is clear that the actual arithmetic is essentially the same but that the results of the former apply to discrete frequencies alone, whereas the results of the latter apply to bands of continuous frequencies. In practice where interest is centred on the spectrum the data will be a sample from a non-periodic process and modification through tapering, addition of zeros, and prewhitening will be necessary for reliable statistical estimates.

Nevertheless, because of the simplicity of harmonic analysis, periodic data will be used initially to describe and explain the two-dimensional spectrum. It will then be a simple step to the interpretation of the results of the two-dimensional Fourier transform.

9.2 Double Fourier series (periodic)

The equation equivalent to Equation (4.4.4) for two dimensions is

$$x[j_1, j_2] = \sum_{k_2 = -n_2/2}^{n_2/2} \sum_{k_1 = -n_1/2}^{n_1/2} c[k_1, k_2] \exp\left\{ i2\pi \left(\frac{j_1 k_1}{n_1} + \frac{j_2 k_2}{n_2} \right) \right\} \quad (9.2.1)$$

or, for a continuous variable,

$$x[t_1, t_2] = \sum_{k_2 = -n_2/2}^{n_2/2} \sum_{k_1 = -n_1/2}^{n_1/2} c[k_1, k_2] \exp\left\{ i2\pi \left(\frac{k_1 t_1}{T_1} + \frac{k_2 t_2}{T_2} \right) \right\} . \quad (9.2.2)$$

The complex coefficients $c[k_1, k_2]$ may be obtained through a similar procedure as outlined in Section 4.5, since

$$\int_{-T_2/2}^{T_2/2} \int_{-T_1/2}^{T_1/2} \exp\left\{i2\pi\left(\frac{k_1 t_1}{T_1} + \frac{k_2 t_2}{T_2}\right)\right\} \exp\left\{-i2\pi\left(\frac{k_1' t_1}{T_1} + \frac{k_2' t_2}{T_2}\right)\right\} dt_1 dt_2$$

$$= 0 \qquad \text{for } k_1 \neq k_1' \text{ and/or } k_2 \neq k_2'$$

$$= T_1 T_2 \qquad \text{for } k_1 = k_1' \text{ and } k_2 = k_2'. \qquad (9.2.3)$$

Therefore

$$c[k_1, k_2] = \frac{1}{T_1 T_2} \int_{-T_2/2}^{T_2/2} \int_{-T_1/2}^{T_1/2} x[t_1, t_2] \exp\left\{-i2\pi\left(\frac{k_1 t_1}{T_1} + \frac{k_2 t_2}{T_2}\right)\right\} dt_1 dt_2$$

$$(9.2.4)$$

or

$$c[k_1, k_2] = \frac{1}{n_1 n_2} \sum_{j_2 = -n_2/2}^{n_2/2} \sum_{j_1 = -n_1/2}^{n_1/2} x[j_1, j_2] \exp\left\{-i2\pi\left(\frac{j_1 k_1}{n_1} + \frac{j_2 k_2}{n_2}\right)\right\} .$$

$$(9.2.5)$$

For computational purposes the latter must be separated into real and imaginary parts by substituting $(a[k_1, k_2] - ib[k_1, k_2])/2$ for $c[k_1, k_2]$ and $\cos\{2\pi(j_1 k_1/n_1 + j_2 k_2/n_2)\} - i\sin\{2\pi(j_1 k_1/n_1 + j_2 k_2/n_2)\}$ for the exponential. Thus Equation (9.2.5) is replaced by two familiar cosine and sine transform equations for the even and odd functions:

$$\frac{a[k_1, k_2]}{2} = \frac{1}{n_1 n_2} \sum_{j_2 = 0}^{n_2 - 1} \sum_{j_1 = 0}^{n_1 - 1} x[j_1, j_2] \cos\left\{2\pi\left(\frac{j_1 k_1}{n_1} + \frac{j_2 k_2}{n_2}\right)\right\} , \qquad (9.2.6)$$

$$\frac{b[k_1, k_2]}{2} = \frac{1}{n_1 n_2} \sum_{j_2 = 0}^{n_2 - 1} \sum_{j_1 = 0}^{n_1 - 1} x[j_1, j_2] \sin\left\{2\pi\left(\frac{j_1 k_1}{n_1} + \frac{j_2 k_2}{n_2}\right)\right\} . \qquad (9.2.7)$$

It should be noted that the summations in the equations are now taken over positive values only of j. Such a change does not alter the result but puts the equation in a form which is more easily applied to the practical situation. Furthermore, it will be seen that the limits of the previous summations are not exact and will vary depending upon whether n_1 and n_2 are even or odd. No restrictions are set on the n's in Equations (9.2.6) and (9.2.7).

That the a's and b's are coefficients or amplitudes of cosine and sine waves should not be difficult to accept, but the interpretation of their coordinates needs some thought. If k_1 and k_2 increase in the same directions as j_1 and j_2, $a[k_1, 0]$ will be found somewhere along the first row of the spectral array. When applied to Equation (9.2.6), $k_2 = 0$ essentially reduces the problem to one dimension. The second row of data $x[j_1, 1]$ is treated as if it continued the first row $x[j_1, 0]$ and the third as if it continued the second, etc. For the coefficient $a[k_1, 0]$ to have magnitude through a least-squares fit there must be a wave of

frequency k_1 in the data. Plotted in the data domain this wave must
crest at j_1, j_2 coordinates, $[0, 0]$, $[0, 1]$, $[0, 2]$, $[0, 3]$, etc., and in other
parallel columns for $k_1 > 1$. In other words, $a[k_1, 0]$ is the amplitude of
a set of parallel straight waves of fixed frequency k_1 lying at right angles
to the direction of $a[k_1, 0]$ from the origin. A similar argument for
$a[0, k_2]$ produces another set of parallel straight waves at right angles to
the direction of that coefficient. The extension of these results leads to
the conclusion that the location of a particular $a[k_1, k_2]$ in the spectral
array provides information on the orientation as well as the frequency of
a set of waves. For example, a simple set with fixed frequency is shown
in Figures 9.2.1 and 9.2.2. There is no phase since the crest occurs at
$0, 0$. Hence the function is even and produces a single spike in the
$a[k_1, k_2]$ spectrum. As there are three complete waves in the basic
interval n_1 and 2 in n_2, the frequencies of 'a' are 3 and 2 (i.e. $a[3, 2]$)
as in Figure 9.2.3. A vector, drawn from the origin to this amplitude, is
perpendicular to the waves. It may be defined by its length, which is
equal to the frequency k_{12} of the waves and by the angle Θ it makes with
the k_1 axis.

$$k_{12} \text{ (per basic interval)} = (k_1{}^2 + k_2{}^2)^{\frac{1}{2}} , \tag{9.2.8}$$

where the basic intervals are equal, i.e. $n_1 \Delta t_1 = n_2 \Delta t_2$. If they are
different,

$$f_{12} \text{ (per units of } t_1 \text{ and } t_2) = \left\{ \left(\frac{k_1}{n_1 \Delta t_1} \right)^2 + \left(\frac{k_2}{n_2 \Delta t_2} \right)^2 \right\}^{\frac{1}{2}} . \tag{9.2.9}$$

The angle is given by

$$\Theta[k_1, k_2] = \arctan \left(\frac{k_2}{n_2 \Delta t_2} \Big/ \frac{k_1}{n_1 \Delta t_1} \right) . \tag{9.2.10}$$

Each of these equations may be developed from the geometry involved in
Figures 9.2.2 and 9.2.3. $k_{12} = (13)^{\frac{1}{2}}$ and $\Theta = 33° 41'$.

Another simple surface shown in Figure 9.2.4 produces two equal
amplitudes lying at right angles to each other in the $a[k_1, k_2]$ spectrum
(Figure 9.2.5). Again, however, the function is even as the waves both
crest at $0, 0$. Where the crests are offset from the origin the $b[k_1, k_2]$
spectrum, obtained from Equation (9.2.7), will have magnitude, and the
phases may be determined from the pairs of a's and b's as in
Equation (2.3.9).

$$\Phi[k_1, k_2] = \arctan \frac{b[k_1, k_2]}{a[k_1, k_2]} \tag{9.2.11}$$

or, in units of t_1 and t_2,

$$\tau_x[k_1, k_2] = \frac{\Phi[k_1, k_2]}{2\pi f_{12}} . \tag{9.2.12}$$

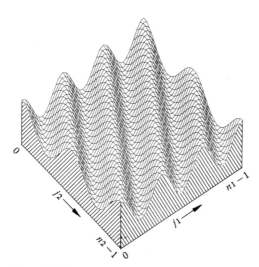

Figure 9.2.1.

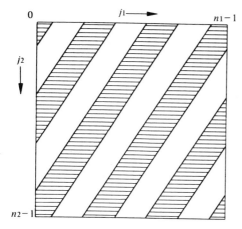

Figure 9.2.2.

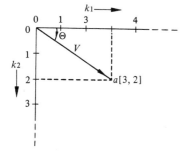

Figure 9.2.3.

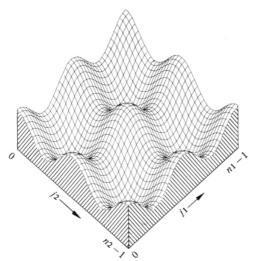

Figure 9.2.4.

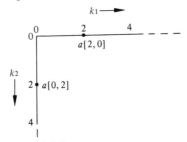

Figure 9.2.5.

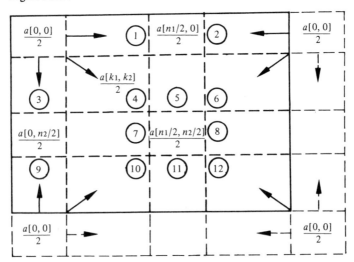

Figure 9.2.6.

Similarly, analogous equations to (2.3.8) and (3.4.3) may be rewritten for the amplitude and variance:

$$A[k_1, k_2] = (a^2[k_1, k_2] + b^2[k_1, k_2])^{\frac{1}{2}} , \qquad (9.2.13)$$

$$\hat{\sigma}^2[k_1, k_2] = \tfrac{1}{2}(a^2[k_1, k_2] + b^2[k_1, k_2]) = \tfrac{1}{2}(A^2[k_1, k_2]) , \qquad (9.2.14)$$

except for $k_1 = k_2 = 0$ when

$$A[0, 0] = \tfrac{1}{2}a[0, 0] \qquad (9.2.15)$$

and the edges when n_1 and/or n_2 are even [see Equations (9.2.16) and (9.2.19)].

So far only waves producing vectors in one quadrant of the spectrum have been considered. Now it is time to look at the whole spectrum. In Figure 4.4.1 and Exercise 4.10.2 it was shown that a one-dimensional variable of n observations would produce a cis spectrum of n terms. Similarly in two dimensions a rectangular array $x[j_1, j_2]$ with $j_1 = j_2 = 0$ at the upper left corner will produce another rectangular array of the same size for the cis spectrum with $k_1 = k_2 = 0$ at the upper left. Separated into $a[k_1, k_2]/2$ and $b[k_1, k_2]/2$ by Equations (9.2.6) and (9.2.7) the arrays must be symmetrical and antisymmetrical respectively. Figure 9.2.6 depicts the arrangement for $a[k_1, k_2]/2$ with both n_1 and n_2 even. As the function is periodic, the first row and the first column are repeated. By analogy to the one-dimensional case the first row $a[k_1, 0]/2$ must be symmetrical about both $[n_1/2, 0]$ and $[0, 0]$. Hence

$$a[1, 0]/2 \equiv a[(n-1), 0]/2, \text{ etc.}$$

A vector drawn from the origin to $[1, 0]$ must point to the right and to $[(n_1-1), 0]$ to the left. In other words, waves perpendicular to j_1 will produce an amplitude both in area ① and in area ②. For the odd function a positive b in ① will appear as a negative b in ② and vice versa. Similarly ③ and ⑨ will be matched. For the other areas matching may be accomplished by pairing parallel vectors. For instance, $a[2, 3]/2$ equals $a[(n_1-2), (n_2-3)]/2$, and the pairs may be listed

① ↔ ②   ⑤ ↔ ⑪

③ ↔ ⑨   ⑥ ↔ ⑩

④ ↔ ⑫   ⑦ ↔ ⑧.

As four origins, as shown in Figure 9.2.6, tend to be confusing it is suggested that the areas be rearranged so as to locate a single origin in the centre as produced by optical analysis (Figure 9.2.7). The k's now take on negative values. An extra row and a column are added for symmetry. Again it should be noted that other coordinate systems are possible. If $x[j_1, j_2]$ had an origin at the bottom left, the coordinates in Figure 9.2.7 would be in the conventional cartesian mode.

Now, since the combination of one-sided a's and b's present all the information on the function, only half this spectrum need be plotted. In the following only the lower half will be calculated. Figure 9.2.8 shows the factors by which Equations (9.2.6) and (9.2.7) must be multiplied to give the one-sided spectra.

If n_1 and/or n_2 are odd the two edge columns and/or bottom row will

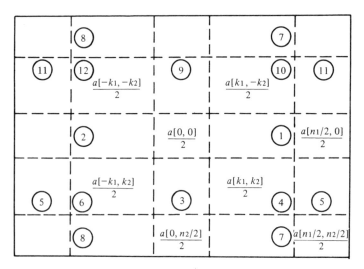

Figure 9.2.7.

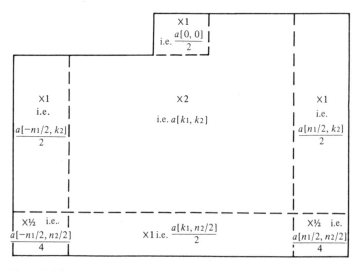

Figure 9.2.8.

be missing. Therefore the modification of Equations (9.2.13) and (9.2.14) for the edges will be, for n's even,

$$A[k_1, k_2] = \frac{(a^2[k_1, k_2] + b^2[k_1, k_2])^{\frac{1}{2}}}{2} \tag{9.2.16}$$

$$\text{for } k_1 = \frac{n_1}{2}, \ k_2 = 0$$

$$k_1 = \frac{n_1}{2}, \ -\frac{n_1}{2}, \ 1 \leqslant k_2 < \frac{n_2}{2}$$

$$-\frac{n_1}{2} < k_1 < \frac{n_1}{2}, \ k_2 = \frac{n_2}{2},$$

$$\hat{\sigma}^2[k_1, k_2] = A^2[k_1, k_2] \text{ for } k\text{'s as in Equation (9.2.16)}, \tag{9.2.17}$$

$$A[k_1, k_2] = \frac{(a^2[k_1, k_2] + b^2[k_1, k_2])^{\frac{1}{2}}}{4} \tag{9.2.18}$$

$$\text{for } k_1 = -\frac{n_1}{2}, \frac{n_1}{2}, \ k_2 = \frac{n_2}{2}$$

$$\hat{\sigma}^2[k_1, k_2] = 2A^2[k_1, k_2] \text{ for } k\text{'s as in Equation (9.2.18)}. \tag{9.2.19}$$

Thus any two-dimensional surface may be represented by a single two-sided two-dimensional discrete spectrum or by a pair of one-sided spectra. The spectral components may be interpreted in exactly the same way as those in the one dimension with the added specification of orientation. The calculation of the coefficients may be performed in a number of ways but the most economical will be by the fast Fourier transform.

9.3 The two-dimensional Fourier integral
9.3.1 The equations
Again by analogy with the one-dimensional case, the Fourier transform may be written

$$X[f_1, f_2] = \int_{-T_2/2}^{T_2/2} \int_{-T_1/2}^{T_1/2} x[t_1, t_2] \exp\{-i2\pi(f_1 t_1 + f_2 t_2)\} \, dt_1 \, dt_2 \ , \tag{9.3.1.1}$$

$$x[t_1, t_2] = \int_{-1/2\Delta t_2}^{1/2\Delta t_2} \int_{-1/2\Delta t_1}^{1/2\Delta t_2} X[f_1, f_2] \exp\{i2\pi(f_1 t_1 + f_2 t_2)\} \, df_1 \, df_2 \ ,$$

$$\tag{9.3.1.2}$$

where $X[f_1, f_2]$ is the amplitude density at the frequency f_1, f_2.

In practice the equations of the previous section will be used to give the total variance which is spread over the band of $(k_1/n_1\Delta t_1) \times (k_2/n_2\Delta t_2)$ centred at k_1, k_2.

9.3.2 Two-dimensional differentiation

For the present purposes it will be assumed that differentiation is required for gradient estimates, say of the ground surface. Interest will be centred upon the magnitude and direction of the maximum slope at any point. As in the one-dimensional case the accuracy of the estimates will depend on the accuracy with which the highest significant frequencies present are recorded.

The method is to calculate the partial derivatives with respect to t_1 and with respect to t_2 and then to combine them to give

$$\text{maximum slope} = \left[\left\{ \frac{\partial(x[t_1, t_2])}{\partial t_1} \right\}^2 + \left\{ \frac{\partial(x[t_1, t_2])}{\partial t_2} \right\}^2 \right]^{\frac{1}{2}} \tag{9.3.2.1}$$

$$\text{direction clockwise from } t_1 = \arctan \left\{ \frac{\partial(x[t_1, t_2])}{\partial t_2} \middle/ \frac{\partial(x[t_1, t_2])}{\partial t_1} \right\} . \tag{9.3.2.2}$$

The partial derivative with respect to t_1 from Equation (9.3.1.2) is given by

$$\frac{\partial(x[t_1, t_2])}{\partial t_1}$$

$$= \int_{-1/2\Delta t_2}^{1/2\Delta t_2} \int_{-1/2\Delta t_1}^{1/2\Delta t_1} \frac{\partial}{\partial t_1} [\![X[f_1, f_2] \exp\{i2\pi(f_1 t_1 + f_2 t_2)\}]\!] df_1 \, df_2$$

$$= \int_{-1/2\Delta t_2}^{1/2\Delta t_2} \int_{-1/2\Delta t_1}^{1/2\Delta t_1} i2\pi f_1 X[f_1, f_2] \exp\{i2\pi(f_1 t_1 + f_2 t_2)\} df_1 \, df_2 . \tag{9.3.2.3}$$

Substituting for $X[f_1, f_2]$ and exp, we find

$$\frac{\partial(x[t_1, t_2])}{\partial t_1}$$

$$= \int_{-1/2\Delta t_2}^{1/2\Delta t_2} \int_{-1/2\Delta t_1}^{1/2\Delta t_1} i2\pi f_1 (a[f_1, f_2] - ib[f_1, f_2])[\cos\{2\pi(f_1 t_1 + f_2 t_2)\}$$

$$+ i\sin\{2\pi(f_1 t_1 + f_2 t_2)\}] df_1 \, df_2$$

$$= \int_{-1/2\Delta t_2}^{1/2\Delta t_2} \int_{-1/2\Delta t_1}^{1/2\Delta t_1} 2\pi f_1 [\![b[f_1, f_2] \cos\{2\pi(f_1 t_1 + f_2 t_2)\}$$

$$\times a[f_1, f_2] \sin\{2\pi(f_1 t_1 + f_2 t_2)\}]\!]$$

$$+ i2\pi f_1 [\![a[f_1, f_2] \cos\{2\pi(f_1 t_1 + f_2 t_2)\}$$

$$- b[f_1, f_2] \sin\{2\pi(f_1 t_1 + f_2 t_2)\}]\!] df_1 \, df_2 . \tag{9.3.2.4}$$

Separation of real and imaginary parts and the writing of discrete coordinates gives

$$\frac{\partial(x[t_1, t_2])}{\partial t_1}[j_1, j_2] = \sum_{k_2=0}^{n_2-1}\sum_{k_1=0}^{n_1-1}\frac{2\pi k_1}{n_1}\left\{b[k_1, k_2]\cos\left(\frac{2\pi j_1 k_1}{n_1} + \frac{2\pi j_2 k_2}{n_2}\right)\right.$$

$$\left. - a[k_1, k_2]\sin\left(\frac{2\pi j_1 k_1}{n_1} + \frac{2\pi j_2 k_2}{n_2}\right)\right\} . \qquad (9.3.2.5)$$

The partial derivative with respect to t_2 is obtained by replacing the multiplication factor $2\pi k_1/n_1$ by $2\pi k_2/n_2$ in Equation (9.3.2.5).

9.4 Convolution and filtering in two dimensions
9.4.1 The mathematical relationships
For the convolution theorem equations equivalent to Equations (5.4.3.1), (5.4.3.2), and (5.4.3.3) are

$$\int_{-\infty}^{\infty}\int_{-\infty}^{\infty}X[f_1, f_2]H[(f_{1_1}-f_1), (f_{2_1}-f_2)]\,df_1\,df_2 \text{ is the Fourier transform of}$$

$$x[t_1, t_2]h[t_1, t_2] , \qquad (9.4.1.1)$$

$X[f_1, f_2]H[f_1, f_2]$ is the Fourier transform of

$$\int_{-\infty}^{\infty}\int_{-\infty}^{\infty}x[t_1, t_2]h[(t_{1_1}-t_1), (t_{2_1}-t_2)]\,dt_1\,dt_2 , \qquad (9.4.1.2)$$

and

$X[f_1, f_2]H^*[f_1, f_2]$ is the Fourier transform of

$$\int_{-\infty}^{\infty}\int_{-\infty}^{\infty}x[(t_{1_1}+t_1), (t_{2_1}+t_2)]h[t_1, t_2]\,dt_1\,dt_2 . \qquad (9.4.1.3)$$

The left-hand side of Equation (9.4.1.1) also may be written $X[f_1, f_2] * H[f_1, f_2]$. Similarly for the discrete calculations Equation (9.4.1.3) becomes [equivalent to Equation (5.4.3.5)]

$$X[k_1, k_2]\,\cancel{n_1 n_2}\,H^*[k_1, k_2]\,n_1 n_2\frac{1}{\cancel{n_1 n_2}}$$

$$= (a_x[k_1, k_2]a_h[k_1, k_2] + b_x[k_1, k_2]b_h[k_1, k_2])\frac{n_1 n_2}{4}$$

$$+ i(a_x[k_1, k_2]b_h[k_1, k_2] - a_h[k_1, k_2]b_x[k_1, k_2])\frac{n_1 n_2}{4} . \qquad (9.4.1.4)$$

These refer to a two-sided function with the a's and b's calculated by Equations (9.2.6) and (9.2.7). They form an $n_1 \times n_2$ matrix as shown, for example, in Figure 9.2.6.

The response of a two-dimensional weighting function $w[t_1, t_2]$
[equivalent to Equation (6.2.13)] is given by

$$W^*[f_1, f_2] = \int_{-\infty}^{\infty} \int_{-\infty}^{\infty} w[t_1, t_2] \exp\{i2\pi(f_1 t_1 + f_2 t_2)\} dt_1 dt_2 . \qquad (9.4.1.5)$$

As in the one-dimensional case this contains two parts, an a_w and a b_w,
from which the phase change may be obtained. If the function is
symmetrical, of course, $b_w[k_1, k_2] = 0$ and the phase change is zero.

In practice the data will be discrete and use must be made of the Dirac
function. Consequently aliasing will occur in two dimensions and will be
just as serious as in the one dimension.

9.4.2 An example of filtering—the calculation of a trend surface

By definition a trend represents only the large scale fluctuations: all the
small-scale oscillations are removed. Frequently its calculation is
accomplished by the fitting of a low-degree polynomial to the empirical
data. Where these data are observed on a rectangular grid, another simple
and obvious method is to use the low frequencies of a harmonic function.
It entails a Fourier transformation, a multiplication by ones at low
frequencies and by zeros at high frequencies, and a retransformation.

As an example a trend surface was calculated from the relief map
shown in Figure 9.4.2.1. Dr. W. Tobler (Tobler, 1968) kindly provided
the original uncorrected digitised data which came from part of the
U.S. Geological Survey, 15-minute series, topographic map for Alma,
Wisconsin. Since the data spacing was relatively large,
$\Delta t_1 = \Delta t_2 = 0·1$ inches at the sheet scale of $1:62500$, some detail was
omitted. The difference may be assessed by comparing the computer
drawn version of the 70×70 digitised array (Figure 9.4.2.1) with the
original topographic map. For the trend calculations this array was first
transformed and then multiplied by a 19-point diameter disc of ones
centred on the mean. A copy of the computer drawn result is shown in
Figure 9.4.2.2. More (less) smoothing would be possible with a smaller
(larger) disc.

Although it is not apparent in this example, large and unwanted
changes may appear in the edges of maps manipulated by this and similar
calculations. The cause is due to the fitting of periodic functions which
must pass smoothly from the values along the bottom immediately to those
along the top and from those along the right edge to those along the left.

In this example it was convenient to define the filtering function
$W^*[f]$ in the spectral domain. Consequently an initial transformation of
the smoothing operator was unnecessary and calculations were
proportionately reduced. A similar simplifying step is not always possible.
For instance, in the pattern recognition example of the next section the
object of search must logically be defined in the data domain and
therefore must be subject to transformation before multiplication.

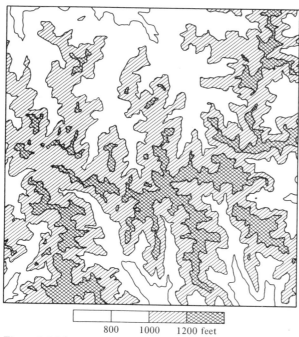

800   1000   1200 feet

Figure 9.4.2.1.

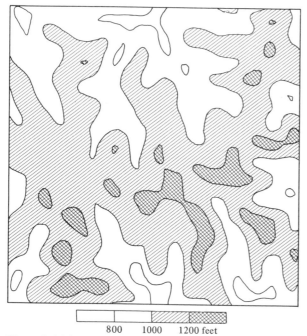

800   1000   1200 feet

Figure 9.4.2.2.

9.4.3 An example of pattern recognition

It was noted in the introduction, Section 1.4.2, that one method of pattern recognition is to move the small known and required pattern, which is often known as the target, over the large and complicated pattern and to calculate the correlation coefficient for each different location.

$$\int_{-\infty}^{\infty}\int_{-\infty}^{\infty} x[(t1_1+t1),(t2_1+t2)]h[t1,t2]\,dt1\,dt2 \; ,$$

where $x[t1,t2]$ is the large array and $h[t1,t2]$ is the target, the object of the search. In fact this is only proportional to the covariance and normalisation with the variances is necessary before the correlation coefficient can be obtained. Nevertheless, use of the covariance alone will demonstrate the simplicity of the technique. From Equation (9.4.1.3) it is clear that the covariance may be obtained through the calculation separately of the Fourier transforms of x and h, their multiplication $X[f1,f2]H^*[f1,f2]$, and a final retransformation.

The above procedure was followed in attempting to find digitally the 'O's in the text in Figure 9.4.3.1. The text was entered by a 66 × 64 array and the target 'O' was placed at the top left-hand corner of another similarly sized array. Location of a portion of a letter was indicated by a constant (e.g. 1) and all other points were designated zero. Final retransformation produced a map of covariances. Only the maxima and second ranked magnitudes are reproduced in the figure. It will be seen that the maxima occur at the top left-hand positions of the letters. This is because the origin of the target pattern was located at its top left corner. Maxima locate not only O's but D's as well. A moment's thought will reveal that this should have been expected. If the covariance had been normalised by the variances of O and D, they would have been separated. Second ranked covariances locate the C, G, and B.

This simple example demonstrates that the technique will work. On the other hand, seldom is the practical problem so straightforward. For instance, it is not always easy to define the pattern sought. The selected target, such as a farmstead, will take on a number of different forms so that it will have to be some statistical representation of all or of a selected set of farmsteads. The resulting $h[t1,t2]$ may well correlate highly with other objects as well as farmsteads. Furthermore, the correlation is highly dependent on orientation, so, if alignment is unknown, as in the case of a search for a particular type of river junction, the target must be rotated and correlations calculated for each angular increment. Even so the technique does reduce the search at the secondary level. Interest is concentrated then on the areas of maximum covariance (or correlation) and the rest of the pattern ignored.

Again it should be emphasised that this is only one method of approach. Related techniques which may make use of the spectrum are edge sharpening by differentiation and alignment filtering. For example, interest may be directed to rivers which do not follow a general alignment. Thus the rivers paralleling the SSW to NNE direction may be removed from Figure 9.5.2.5 by a filter, $H^*[f_1, f_2]$, which contains zeros in wedges in the WNW and ESE directions. Examples of alignment filtering are to be found in Dobrin *et al.* (1965), Pincus and Dobrin (1966), and Bauer *et al.* (1967).

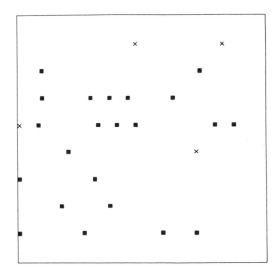

Figure 9.4.3.1.

9.5 Spectral analysis in two dimensions

9.5.1 The steps involved in using the direct method

The two-dimensional spectrum may be calculated either by the use of the modified covariance array as discussed by Pierson (1960) and by Leese and Epstein (1963) or by the direct method using the fast Fourier transform technique. The same arguments that applied to the one-dimensional case apply to the two-dimensional case. Since the raw sample data are observed through a window, which has undesirable properties, it must be replaced by a more tractable one. In order to increase the degrees of freedom and hence the confidence in the results the Fourier coefficients $a[k_1, k_2]$ and $b[k_1, k_2]$, which produce the elementary variance bands, must be combined into groups, and to minimise leakage between bands the mean and trends should be removed and possibly prewhitening performed. The steps may be summarised for a $D_1 \times D_2$ array of $x'[j_1, j_2]$.

Step 1 remove mean and trends.

Step 2 prewhiten if necessary with

$$x[j_1, j_2] = x'[j_1, j_2] - E_1(x'[j_1+1, j_2] + x'[j_1-1, j_2])$$
$$- E_2(x'[j_1, j_2+1] + x'[j_1, j_2-1]), \qquad (9.5.1.1)$$

where $0 < E_1 < 1$, $0 < E_2 < 1$ are the prewhitening constants, the larger the E the larger the effect on the spectrum. The high frequencies are magnified relative to the low.

Step 3 taper with the window function $h[j_1, j_2]$ to give

$$\hat{x}[j_1, j_2] = x[j_1, j_2] h[j_1, j_2], \qquad (9.5.1.2)$$

where

$$h[j_1, j_2] = \tfrac{1}{2}\left\{1 - \cos\left(\frac{\pi j_1}{G_1}\right)\right\} \qquad 0 \leqslant j_1 \leqslant G_1$$

$$= \tfrac{1}{2}\left\{1 - \cos\left(\frac{\pi j_2}{G_2}\right)\right\} \qquad 0 \leqslant j_2 \leqslant G_2$$

$$= 1 \qquad G_1 < j_1 < D_1 - G_1, G_2 < j_2 < D_2 - G_2$$

$$= \tfrac{1}{2}\left[1 - \cos\left\{\frac{\pi(D_1 - j_1)}{G_1}\right\}\right] \qquad D_1 - G_1 \leqslant j_1 < D_1$$

$$= \tfrac{1}{2}\left[1 - \cos\left\{\frac{\pi(D_2 - j_2)}{G_2}\right\}\right] \qquad D_2 - G_2 \leqslant j_2 < D_2 \quad (9.5.1.3)$$

and where G_1 and G_2 are the numbers of columns and rows tapered. In the first analysis Tukey (1967) suggests they should be approximately 10% of D_1 and D_2. Subsequently other windows should also be tested.

Step 4 add zeros for the beginning and/or end columns and rows so that the array $\hat{x}[j_1, j_2]$ has dimensions $n_1 \times n_2$.

Step 5 calculate the two-dimensional Fourier transform coefficients. For discrete data the two-sided elementary bands are given by $\hat{a}[k_1, k_2]/2$ and $\hat{b}[k_1, k_2]/2$ of Equations (9.2.6) and (9.2.7). In practice the fast Fourier transform algorithm should be used to produce the $\hat{a}$'s and $\hat{b}$'s from the $\hat{x}[j_1, j_2]$.

Step 6 post-darkening must accompany prewhitening.

Step 7 calculate the variances by summing. Here the summing is done within rectangles $(2z_1 + 1) \times (2z_2 + 1)$ but other shaped areas may be more useful (see Section 9.5.2):

$$\hat{XX}[r_1, r_2] = \sum_{k_2 = r_2(2z_2 + 1) - z_2}^{r_2(2z_2 + 1) + z_2} \sum_{k_1 = r_1(2z_1 + 1) - z_1}^{r_1(2z_1 + 1) + z_1} \frac{\hat{a}^2[k_1, k_2] + \hat{b}^2[k_1, k_2]}{2}$$

$$\text{for } 0 < r_1 < m_1, \; 0 < r_2 < m_2 . \qquad (9.5.1.4)$$

For the edge rows and columns, which will be $z_1 + 1$ and $z_2 + 1$ wide respectively, the variances of edge elementary rows and columns are given by Equations (9.2.16) to (9.2.19). Schematically this is shown in Figure 9.5.1.1 for $z_1 = z_2 = 1$, $m_1 = m_2 = 3$.

Thus, for example,

$$\hat{XX}[0, 0] = \hat{a}^2[0, 0] + \sum_{k_1 = 1}^{z_1} \frac{\hat{a}^2[k_1, 0] + \hat{b}^2[k_1, 0]}{2}$$

$$+ \sum_{k_2 = -1}^{z_2} \sum_{k_1 = -z_1}^{z_1} \frac{a^2[k_1, k_2] + b^2[k_1, k_2]}{2} . \qquad (9.5.1.5)$$

The resulting estimates will be located at frequencies $[r_1/2m_1\Delta t_1, r_2/2m_2\Delta t_2]$ where $0 \leqslant r_1 \leqslant m_1$, $0 \leqslant r_2 \leqslant m_2$. As in the one-dimensional case, n_1 need not be even and $(n_1/2)/m_1$ need not equal $2z_1 + 1$ and be odd; similarly

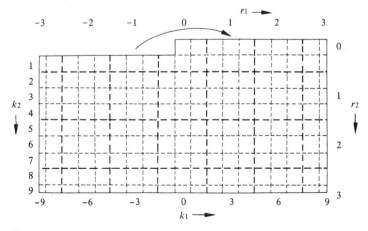

Figure 9.5.1.1.

for n_2. These particular relationships were chosen because final summation blocks then have a one-to-one relationship with the elementary blocks.

Confidence intervals for $\hat{X}\hat{X}[r_1, r_2]$ will still be derived from χ^2, where degrees of freedom are given by

$$\nu_{12} = \nu_1 \times \nu_2$$

$$\approx \frac{(2z_1 + 1)}{n_1/2}(D_1 - G_1) \times \frac{(2z_2 + 1)}{n_2/2}(D_2 - G_2) \qquad (9.5.1.6)$$

except for $r_1 = 0$, $r_2 = 0, m_2$; $r_1 = \pm m$, $0 < r_2 < m_2$; $-m_1 < r_1 < m_1$, $r_2 = m_2$; which contain half that number of degrees of freedom and $r_1 = \pm m$, $r_2 = m_2$ which contains a quarter of the number given by Equation (9.5.1.6).

9.5.2 Examples of two-dimensional spectral analysis

Two types of drainage pattern as shown in Figures 9.5.2.1 and 9.5.2.5 have been analysed using the steps listed in Section 9.5.1. The resulting one-sided variance spectra are reproduced in Figures 9.5.2.2 and 9.5.2.6.

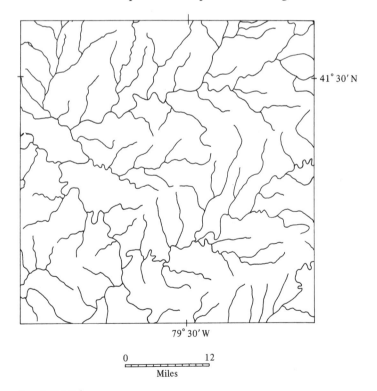

41° 30' N

79° 30' W

0 _____ 12
Miles

Figure 9.5.2.1.

The original data at one-twentieth of an inch grid were read off the
Commonwealth of Pennsylvania Department of Forests and Waters stream
map published in 1930 at a scale of 1 : 380 160. In each case this
produced 140 x 140 arrays from the seven-inch square areas, the locations
of which are shown by latitude and longitude. The positions of the
streams were indicated by ones in a background of zeros. Consequently,
the absolute values of the variance spectra, which have been multiplied by
10^6, have little meaning, and interest is directed to the variations within
the spectra. Because the observational technique registered only location,
the variance spectra indicate the relative spacing and alignment of the
rivers and contain no information on width or volume of flow.
Alternative methods of representing rivers must be chosen if the latter
characteristics are required. For example, width might be obtained from
aerial photographs, and volume might be estimated from the fitting of
smooth profile curves to gauging station data. The full computational
factors chosen for both patterns were $D_1 = D_2 = 140$, $n_1 = n_2 = 200$,
$G_1 = G_2 = 14$, and $m_1 = m_2 = 21$, leading to the same frequency
intervals for each direction.

The dendritic pattern, Figure 9.5.2.1, as to be expected, shows no
distinct orientation. Maximum variance values occur in the first half-
dozen or so frequency bands with clear minima at the origin (very small
frequencies—large distances between the streams) and at large frequencies
(small distances between streams). If orientation is unimportant, the
two-dimensional spectrum is not needed and it may be collapsed to one
dimension. This may be accomplished by summing the variances in

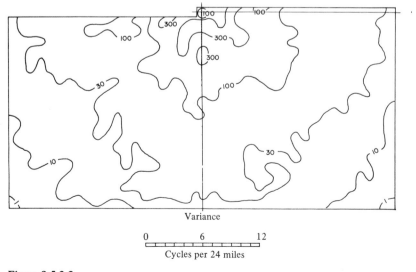

Variance

Figure 9.5.2.2.

semicircular strips of constant frequency (see Figure 9.5.2.3). In the
present example the final blocks plotted in Figure 9.5.2.2 were used but
the elementary blocks may be manipulated in the same way. It should
be noted that as frequency increases the number of elementary blocks
summed, and therefore the number of degrees of freedom, increases.
The spectrum of dendritic streams with orientation removed is presented
in Figure 9.5.2.4. It shows a rapid increase in variance from the origin
to a frequency of 0·167 cycles per mile (6-mile spacing) and then a
gradual decrease. Amongst other things it gives a measure of the average
distances independent of direction between streams. The most likely
distance appears to be 6 miles but smaller distances are also present.

The pattern in Figure 9.5.2.5 is clearly quite different with distinct
alignments, a fact which is born out by the spectrum in Figure 9.5.2.6.
The apportioned variance still increases from the origin but only along
a WNW to ESE direction, that is at right angles to the waves or, in this
case, streams. There is no need to collapse the spectrum to the non-
oriented one dimension and indeed it would only confuse the information.
The majority of streams run parallel to one another at spacings between
three and six miles, a similar scale to that obtained for an unaligned
dendritic pattern in the same general region. In other examples, where

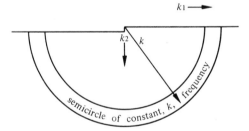

Figure 9.5.2.3.

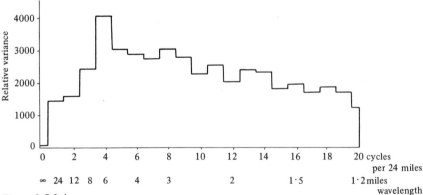

Figure 9.5.2.4.

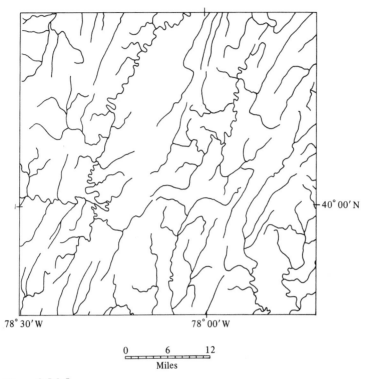

Figure 9.5.2.5.

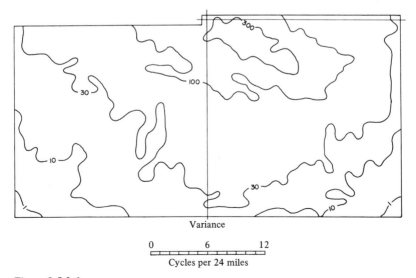

Figure 9.5.2.6.

alignments are not so clearly defined, methods are available for calculating the best directional angle and also the angular width of the spectrum (Longuet-Higgins, 1957).

It might be argued that no information has been gained and that these results could have been predicted. Such a statement is not altogether true. The average distances and orientations have been quantified, a step which is not easily performed by other techniques, yet one which is necessary in many phases of research relating to stream patterns. Furthermore, the success and usefulness of the technique in these two simple examples suggest that it may be applied in more complicated situations where alignments or non-alignments are not so obvious. Again, the analysis related solely to a pattern and other stream characteristics such as volume or rate of flow would produce proportionately more information. Of course, these comments apply only to stream analysis. The value of two-dimensional spectral analysis has already been demonstrated in other fields such as oceanography (see, for example, N.A.S., 1963).

9.6 Cross spectral analysis in two dimensions

9.6.1 The equations

As in many facets of two-dimensional analysis the same arguments as were used for the one dimension hold for two and need not be repeated. The cross spectra may be derived from the two-dimensional lag covariances or directly from the original data arrays. If the direct method is followed, preliminary to cross spectral analysis all the steps involved in calculating the a's and b's of the elementary bands or blocks of the one-variable two-dimensional spectrum must be followed. For two variables $x[j1, j2]$ and $y[j1, j2]$, step 6 of Section 9.5.1 produces $\hat{a}_x[k1, k2]$, $\hat{b}_x[k1, k2]$, $\hat{a}_y[k1, k2]$, and $\hat{b}_y[k1, k2]$. These are then used as input to the two-dimensional equivalent equations of Sections 8.4 and 8.3. The first two read for rectangularly summed blocks (other shaped areas may be used): the *one-sided* cospectrum

$$\hat{XY}_E[r1, r2] = \sum_{k2=r2(2z2+1)-z2}^{r2(2z2+1)+z2} \sum_{k1=r1(2z1+1)-z1}^{r1(2z1+1)+z1} \left(\frac{\hat{a}_x[k1, k2]\hat{a}_y[k1, k2]}{2} \right. $$
$$\left. + \frac{\hat{b}_x[k1, k2]\hat{b}_y[k1, k2]}{2} \right) \qquad (9.6.1.1)$$

and the *one-sided* quadrature spectrum

$$\hat{XY}_O[r1, r2] = \sum_{k2=r2(2z2+1)-z2}^{r2(2z2+1)+z2} \sum_{k1=r1(2z1+1)-z1}^{r1(2z1+1)+z1} \left(\frac{\hat{a}_x[k1, k2]\hat{b}_y[k1, k2]}{2} \right. $$
$$\left. - \frac{\hat{a}_y[k1, k2]\hat{b}_x[k1, k2]}{2} \right) . \qquad (9.6.1.2)$$

Edge rows and columns of the cospectrum and quadrature spectrum, like the variance spectrum, require special attention and modification to these formulae (see Exercise 9.7.2). The cross covariance is given by

$$| \hat{XY}[r_1, r_2] | = (\hat{XY}_E^2[r_1, r_2] + \hat{XY}_O^2[r_1, r_2])^{\frac{1}{2}} \qquad (9.6.1.3)$$

and the phase by

$$\Phi_{yx}[r_1, r_2] = \arctan\left(\frac{\hat{XY}_O[r_1, r_2]}{\hat{XY}_E[r_1, r_2]}\right), \qquad (9.6.1.4)$$

where the first crest of a parallel set of waves of frequency $[r_1, r_2]$ in array $y[j_1, j_2]$ lies further from the origin than a similar crest in array $x[j_1, j_2]$ by the phase shift $\Phi_{yx}[r_1, r_2]/(r_1^2 + r_2^2)^{\frac{1}{2}}$. Or, in the units of t_1 and t_2 this distance may be expressed by the lag

$$\tau_{yx}[r_1, r_2] = \frac{\Phi_{yx}[r_1, r_2]/2\pi}{\{(r_1/2m_1\Delta t_1)^2 + (r_2/2m_2\Delta t_2)^2\}^{\frac{1}{2}}} . \qquad (9.6.1.5)$$

The proportionate increase in the parallel waves of y for a given x set at the same frequency is estimated by the gain

$$\hat{V}_{yx}[r_1, r_2] = \frac{| \hat{XY}[r_1, r_2] |}{\hat{XX}[r_1, r_2]} \qquad (9.6.1.6)$$

or

$$\hat{V}_{xy}[r_1, r_2] = \frac{| \hat{XY}[r_1, r_2] |}{\hat{YY}[r_1, r_2]} \qquad (9.6.1.7)$$

and the correlation or coherence becomes

$$\hat{R}_{yx}^2[r_1, r_2] = \frac{\hat{XY}_E^2[r_1, r_2] + \hat{XY}_O^2[r_1, r_2]}{\hat{XX}[r_1, r_2]\, \hat{YY}[r_1, r_2]} . \qquad (9.6.1.8)$$

9.6.2 An example

Of the vast number of pairs of patterns which might have been chosen for this example, rivers and relief were selected because a visual correlation is usually present and because physical processes connect the two. To match the river pattern of Figure 9.5.2.5 the relief array, for approximately the same grid points, was read off the U.S.G.S. 1 : 250000 scale topographic maps, Pittsburgh, Harrisburg, Cumberland, and Baltimore. Although the original contour interval was 100 feet, to speed the reading of the 20000 points, the data were interpolated from the 500 foot contours shown in Figure 9.6.2.1. This has the effect of removing much of the fine detail which would appear mainly in the high frequencies. Use of the same parameters as for the examples in Section 9.5.2 produces the spectrum shown in Figure 9.6.2.2. Despite a clear slope from NW to SE no trend removal was performed so high variances occur around $[0, 0]$. In a more detailed analysis this trend should be filtered out as it may affect all frequencies. Hence the present results are tentative. The same

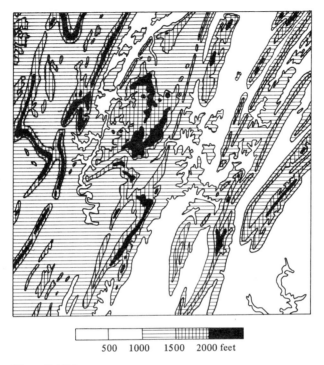

500   1000   1500   2000 feet

Figure 9.6.2.1.

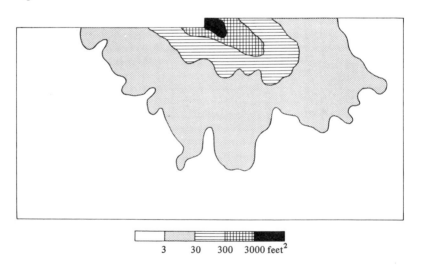

3   30   300   3000 feet^2

Figure 9.6.2.2.

NNE-to-SSW alignment as observed in original relief map is evident, although it is not as distinct as the one displayed in the corresponding river pattern.

The simple correlation between the two tapered arrays is $-0 \cdot 1$. This might have been expected from the raw data. The river pattern, composed of ones and zeros, contains little variability in comparison with the relief. All rivers are represented as uniform maxima, whereas the corresponding points in the relief array are minima of varying magnitudes. Similarly the interfluves of the river pattern are uniform minima, whereas the same areas in the relief map are highly variable.

The coherence spectrum $\hat{R}^2_{yx}[r_1, r_2]$, drawn for values of $0 \cdot 4$ or greater in Figure 9.6.2.3, reveals no outstanding relationships. The largest estimate of $0 \cdot 6$ (90% confidence limits of $0 \cdot 3$ and $0 \cdot 8$), at $k_1 = 2$, $k_2 = 1$, lies in the same general alignment region as the variance maxima of the individual spectra but there is no particular concentration of coherence. The coherence, R^2, may be interpreted in a similar way to the coefficient of determination in simple linear regression. A coherence of $0 \cdot 6$ indicates that statistically in the sample for a specific frequency block some 60% of the variance in one variable may be accounted for by the other. Despite the generally low coherence and wide confidence bands the phase does provide some interesting information. At $k_1 = 2$, $k_2 = 1$ the phase is $1 \cdot 7$ radians, which indicates that along this alignment the maximum of the relief is further from the origin by $1 \cdot 7/6 \cdot 28$ cycle than the rivers. In other words, the rivers at these frequencies tend to lie closer to the ridges on the ESE than those on the WNW. Other phase estimates in the same alignment region of the spectrum bear this out and it is supported by an inspection of the original maps. Of course, such a conclusion is tentative and further analysis is necessary but it does suggest one possible application of the technique. Note that no knowledge of the underlying geological structure was presupposed.

Because of the nature of the river input data the 'gain' has little significance in the present study. Had volume or rate of flow been used, possible physical meaning might have been attached to this parameter.

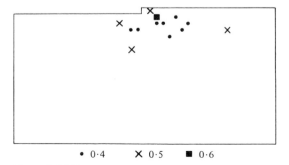

$\bullet$ $0 \cdot 4$ $\times$ $0 \cdot 5$ $\blacksquare$ $0 \cdot 6$

Figure 9.6.2.3.

9.7 Exercises

9.7.1 For the following array calculate the $a[k_1, k_2]$'s and $b[k_1, k_2]$'s assuming that the array is periodic.

4	5	2	5
2	6	3	2
3	8	1	3
4	7	4	6

(1) Write out the arrays for a and b (a) as in Figure 9.2.6; (b) as in Figure 9.2.8.
(2) Write out the array for A as in Figure 9.2.8.
(3) Write out the array for $\hat{\sigma}^2$ as in Figure 9.2.8.

9.7.2 Write the equation for $\hat{X}X[m_1, m_2]$ equivalent to Equation (9.5.1.4) in terms of the a's and b's. Note that step 5 of Section 9.5.1 uses Equations (9.2.6) and (9.2.7).

9.7.3 If it were necessary to standardise the covariances calculated in Section 9.4.3.1 (i.e. to produce correlation coefficients), what would be an obvious spectral method of calculating the standard deviations required?

9.7.4 What will be the difference in the results of convolving the data in Exercise 9.7.1 separately with the following two arrays?

0·25	0·25	0	0	0	0	0	0
0·25	0·25	0	0	0	0·25	0·25	0
0	0	0	0	0	0·25	0·25	0
0	0	0	0	0	0	0	0

9.7.5 Reading: Harbaugh and Preston (1965)—note that they calculate four coefficients instead of a and b; Leese and Epstein (1963); N.A.S. (1963); Pierson (1960).

10

Concluding remarks

10.1 Optical techniques.

All the basic procedures of Chapter 9 may be performed optically and each set of equations has an optical analogue. The growth of the optical approach began with the introduction of the laser which had the advantage of producing both coherence and intense light. Non-coherent light sources may be used but some information relating to phase is lost. The general arrangement of a typical optical bench is shown in Figure 10.1.1.

The laser light issuing from the collimator lens is essentially parallel. On passing through the image transparency (data $x[j_1, j_2]$) the light is diffracted, only to be concentrated again by the transform lens. The distribution of the light at the focal plane (Fourier transform plane) turns out to be the two-dimensional spectrum of the original image with the mean at the centre. If some kind of sensor such as a photographic plate or optical digitiser is inserted at the transform plane, then the two-dimensional Fourier transform will be recorded.

If filtering is required, the transform of the filter may be inserted at this point. For instance, in the example of trend calculation in Section 9.4.2 the 19-point diameter disc turns out to be a circular hole of proportionately the same diameter. The light from the original image of the topography is thus selectively passed. Another transform lens will convert the modified spectrum back to an image which is the required smoothed topography. The advantages of the optical approach are clearly evident if the amount of smoothing required is unknown. The operator, using a variable aperture in the transform plane, can change the degree of smoothing at will and observe the results simultaneously. The same holds true for pattern recognition where perhaps orientation is crucial. The target pattern may be set in a revolving mount at the transform plane and the varying covariance surface observed and if necessary recorded.

For the statistical two-dimensional spectrum, tapering involves fading out the edges of the original image and final summation into blocks or areas depends upon the resolution of the recording device. For instance, a coarse flying spot scanner will sum over large areas.

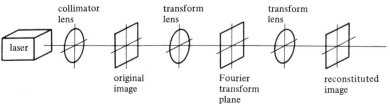

Figure 10.1.1

Another advantage of the optical technique is its immediate application to data which are in photographic forms. On the other hand, photographs from satellites are basically digitised arrays. Thus they are more applicable to digital analysis.

10.2 Review

In the previous chapters an attempt has been made to explain the spectrum and its uses to the student who has little mathematics and knows nothing about the Fourier transform. For those with a firmer base in the subject the exposition of the actual techniques and the discussion of their wide range of application should have proved useful. A convenient classification scheme has been the division between those procedures used in estimating the statistical spectrum, where a scale decomposition is required as an end product, and those where Fourier transformations simplify computation. The difference between the one- and two-dimensional analysis is less significant although, in the initial development, the one-dimensional formulae and ideas are easier to comprehend. Similarly, two variable manipulations in the cross spectra follow naturally from simple regression concepts.

The aim of this monograph was not to give an exhaustive review of the techniques. Many facets have been ignored (e.g. complex demodulation) and new approaches and discoveries are modifying existing views. Rather, the intention was to create a base from which a student might branch out on his own and read original theoretical and applied papers with understanding. As in any area, the key to real understanding is the ability to apply and to test theoretical concepts.

In each section the equations have been given in an explicit form which should allow the reader to make his own calculations. In practice this means the writing of the equations in computer language and hence presupposes that a computer is available. The short exercises should have convinced the user that the range of application of any of the techniques mentioned is severely limited if a computer is not accessible.

It has been intimated that the subject is a dynamic one and that the future will reveal new developments and new areas of application. Already it has been demonstrated that the general technique has a place in multivariate analysis (Jenkins and Watts, 1968, chapter 11) and in the study of higher order moments of a series (Hasselmann et al., 1963). Three-dimensional analyses have been applied in crystallography but apparently not legitimately (Donohue, 1969). In each of these areas the preliminary results indicate that Fourier techniques may be one set of keys for unlocking large stores of information which are presently unintelligible. Again it cannot be overemphasised that quantitative techniques of this type are nothing more than tools which aid in man's quest for knowledge.

References

These references have been found useful by the author. Only those marked ● have been referred to in the text. For a partial classification of these, reference should be made to Appendix B.

Agterberg, F. P., 1967, "Computer techniques in geology", *Earth Sci. Rev.*, **3**, 47-77.

Akaike, H., 1967, "Some problems in the application of the cross-spectral method", in *Advanced Seminar on Spectral Analysis of Time Series*, Ed. B. Harris (Wiley, New York), pp.81-107.

Aki, K., 1955, "Correlogram analysis of seismograms", *Geophys. Notes, Tokyo*, **8**, 99-107.

Alavi, A. S., Jenkins, G. M., 1965, "An example of digital filtering", *Appl. Statist.*, **14**, 70-74.

Aller, L. H., 1951, "Turbulence in the interstella medium", *Astrophys. J.*, **113**, 120-124.

Alsop, L. E., Nowroozi, A. A., 1966, "Fast Fourier analysis", *J. geophys. Res.*, **71**, 5482-5483.

Anderson, R. L., 1942, "Distribution of the serial correlation coefficient", *Ann. math. Statist.*, **13**, 1-13.

Anderson, R. Y., Koopmans, L. H., 1963, "Harmonic analysis of varve time series", *J. geophys. Res.*, **68**, 877-893.

Angell, J., Korshover, J., 1963, "Harmonic analysis of the biennial zonal wind and temperature regimes", *Mon. Weath. Rev. U.S. Dep. Agric.*, **91**, 537-548.

Barber, N. F., 1954, "Finding the direction of travel of sea waves", *Nature*, **174**, 1048-1050.

●Barber, N. F., 1961, *Experimental Correlograms and Fourier Transforms* (Pergamon Press, London).

●Barber, N. F., 1966, "Fourier methods in geophysics", in *Methods and Techniques in Geophysics*, **2**, Ed. S. K. Runcorn (Wiley, New York), pp.123-204.

Barnard, G. A., Jenkins, G. M., Winston, C. B., 1962, "Likelihood, inference and time series", *Jl R. statist. Soc., Ser. A*, **125**, 321-372.

Bartels, J., 1935, "Random fluctuations, persistence, and quasi persistence in geophysical and cosmical periodicities", *Terr. Magn. atmos. Elect.*, **40**, 1-60.

Bartlett, M. G., 1946, "On the theoretical specification and sampling properties of auto-correlated time series", *Jl R. statist. Soc.*, supplement 8, 27-41.

Bartlett, M. G., 1948, "Smoothing periodograms from time series with continuous spectra", *Nature*, **161**, 686-687.

●Bartlett, M. G., 1950, "Periodogram analysis and continuous spectra", *Biometrika*, **37**, 1-16.

Bartlett, M. G., 1963, "The spectral analysis of point processes", *Jl R. statist. Soc., Ser. B*, **25**, 264-296.

Bartlett, M. G., 1964, "The spectral analysis of two-dimensional point processes", *Biometrika*, **51**, 299-311.

Bartlett, M. S., Mehdi, J., 1955, "On the efficiency of procedures for smoothing periodograms from time series with continuous spectra", *Biometrika*, **42**, 143-150.

●Bauer, A., Fontanel, A., Grau, G., 1967, "The application of optical filtering in coherent light to the study of aerial photographs of Greenland glaciers", *J. Glaciol.*, **6**, 781-793.

Bendat, J. S., 1962, "Interpretation and application of statistical analysis for random physical phenomena", *I.R.E. Trans. med. Electron.*, BME-9, 31-43.

●Bendat, J. S., Piersol, A. G., 1966, *Measurement and Analysis of Random Data* (Wiley, New York).

Bennett, J., Carter, V., Kuratani, Y., 1968, "The business cycle in price series—a spectral view", Research Report, Batelle Memorial Institute, Columbus, 43201.

●Beveridge, W. H., 1921, "Weather and harvest cycles", *Econ. J.*, **31**, 429-452.
●Beveridge, W. H., 1922, "Wheat prices and rainfall in Western Europe", *Jl R. statist. Soc.*, **85**, 412-478.
Bingham, C., Godfrey, M. D., Tukey, J. W., 1967, "Modern techniques of power spectrum estimation", *I.E.E.E. Trans. Audio Electroacoust.*, AU-15, number 2, 56-66.
●Blackman, R. B., Tukey, J. W., 1958, "The measurement of power spectra", *Bell Syst. tech. J.*, **37**, 185-282, 485-569, also as a book 1959 (Dover Publications, New York).
Bogert, B. P., Healey, M. J., Tukey, J. W., 1963, "The quefrency alanysis of time series for echoes: cepstrum, pseudo-autocovariance, cross-cepstrum, and saphe cracking", in *Proceedings of the Symposium on Time Series Analysis*, Ed. M. Rosenblatt (Wiley, New York), pp.201-243.
●Bogert, B. P., Parzen, E., 1967, "Informal comments on the uses of power spectrum analysis", *I.E.E.E. Trans. Audio Electroacoust.*, AU-15, 74-76.
Born, B. P., Wolf, E., 1959, *Principle of Optics* (Pergamon Press, Oxford).
Boville, B. W., Kwizak, M., 1959, "Fourier analysis applied to hemispheric waves of the atmosphere", Tech. Circular number CIR-3155 TEC-292, Met. Branch, Dept. Transport, Canada.
Box, G. E. P., 1960, "Some general considerations in process optimisation", *Trans. Am. Soc. mech. Engrs.*, **82**, 113-119.
Box, G. E. P., Jenkins, G. M., Bacon, D. W., 1967, "Models for forecasting seasonal and non-seasonal time series", in *Advanced Seminar on Spectral Analysis of Time Series*, Ed. B. Harris (Wiley, New York), pp.271-311.
Brillinger, D. R., 1965, "An introduction to polyspectra", *Ann. math. Statist.*, **36**, 1351-1374.
Brillinger, D. R., Rosenblatt, M., 1967a, "Asymptotic theory of estimates of kth order spectra", in *Advanced Seminar on Spectral Analysis of Time Series*, Ed. B. Harris (Wiley, New York), pp.153-188.
Brillinger, D. R., Rosenblatt, M., 1967b, "Computation and interpretation of kth order spectra", *ibid.*, pp.189-233.
Brooks, C. E. P., Carruthers, N., 1953, *Handbook of Statistical Methods in Meteorology* (HMSO, London).
Brown, J. A., 1964, "A diagnostic study of tropospheric diabatic heating and the generation of available potential energy", *Tellus*, **16**, 371-388.
●Brownlee, J., 1917, "An investigation into the periodicity of measles epidemics in London from 1703 to the present day by the method of periodogram", *Phil. Trans. R. Soc., Ser. B*, **208**, 225-250.
●Bryson, R. A., Dutton, J. A., 1961, "Some aspects of the variance spectra of tree rings and varves", *Ann. N.Y. Acad. Sci.*, **95**, 580-604.
Bryson, R. A., Dutton, J. A., 1967, "The variance spectra of certain natural series", in *Quantitative Geography*, Eds. W. Garrison, D. F. Marble, Northwestern University Studies in Geography, number 13 (Northwestern University, Evanston), pp.1-24.
Busch, N. E., Panofsky, H. A., 1968, "Recent spectra of atmospheric turbulence", *Q. Jl R. met. Soc.*, **94**, 132-148.
Buys-Ballot, C. H. D., 1847, *Les Changements périodiques de Température* (Kemink and Zoon, Utrecht).
Byerly, P. E., 1965, "Convolution filtering of gravity and magnetic maps", *Geophysics*, **30**, 281-284.
Carslaw, H. S., 1930, *Introduction to the Theory of Fourier Series and Integrals* (Macmillan, London).
Carson, J. E., 1963, "Analysis of soil and air temperatures by Fourier techniques", *J. geophys. Res.*, **68**, 2217-2232.

Cartwright, D. E., Rydill, L. J., 1957, "The rolling and pitching of a ship at sea: a direct comparison between calculated and recorded motion of a ship in sea waves", *Trans. Instn. nav. Archit.*, **99**, 100-135.

Casetti, E., 1966, "Analysis of spatial association by trigonometric polynomials", *Can. Geogr.*, **10**, 199-204.

Chapman, S., 1969, "The lunar and solar semidiurnal variations of barometric pressure at Copenhagen 1884-1949 (66 years)", *Q. Jl. R. met. Soc.*, **95**, 381-394.

●Chapman, S., Bartels, J., 1940, "Periodicity and harmonic analysis", chapter 16 in *Geomagnetism*, **II**, *Analysis of the Data and Physical Theories* (Oxford University Press, Oxford), pp.543-603.

Charney, J. G., 1947, "The dynamics of long waves in a baroclinic westerly current", *J. Met.*, **4**, 135-162.

Charnock, H., 1957, "Notes on the specification of atmospheric turbulence", *Jl R. statist. Soc., Ser. A*, **120**, 398-425.

●Cheng, G. C., Ledley, R. S., Pollock, D. K., Rosenfeld, A., (Eds.), 1968, "Pictorial pattern recognition", *Proceedings of a Symposium on Automatic Photointerpretation held in Washington in 1967* (Thompson Book Co., Washington).

Chiu, Wan-chang, 1960, "The spectra of large scale turbulent transfer of momentum and heat", *J. Met.*, **17**, 435-441.

●Clayton, H. H., 1917, "Effect of short period variation of solar radiation on the Earth's atmosphere", *Smithson. misc. Collns.*, **68**, 1-76.

Cochran, W. T., Cooley, J. W., Favin, D. L., Helms, H. D., Kaenel, R. A., Lang, W. W., Maling, G. C. Jr., Nelson, D. E., Rader, C. M., Welch, P. D., 1967, "What is the fast Fourier transform?", *I.E.E.E. Trans. Audio Electroacoust.*, AU-15, number 2, 45-55.

●Conrad, V., Pollack, L. W., 1950, *Methods in Climatology* (Harvard University Press, Cambridge, Mass.).

Cooley, J. W., 1966, "Applications of the fast Fourier transform method", in *Proceedings of the IBM Scientific Computing Symposium on the Digital Simulation of Continuous Systems, 1966* (IBM, White Plains, New York).

●Cooley, J. W., Tukey, J. W., 1965, "An algorithm for machine calculation of complex Fourier series", *Maths. Comput.*, **19**, 297-301.

●Cooley, J. W., Lewis, P. A. W., Welch, P. D., 1967a, "Historical notes on the fast Fourier transform", *I.E.E.E. Trans. Audio Electroacoust.*, AU-15, number 2, 76-79.

●Cooley, J. W., Lewis, P. A. W., Welch, P. D., 1967b, "Application of the fast Fourier transform to computation of Fourier integrals, Fourier series, and convolution integrals", *I.E.E.E. Trans. Audio Electroacoust.*, AU-15, number 2, 79-84.

Cornfield, J., Tukey, J. W., 1956, "Average values of mean squares in factorials", *Ann. math. Statist.*, **27**, 907-949.

Cox, C., Munk, W., 1954, "Statistics of the sea surface derived from sun glitter", *J. mar. Res.*, **13**, 198-227.

Cox, D. R., Lewis, P. A. W., 1966, *The Statistical Analysis of Series of Events* (Methuen Monographs, London; Wiley, New York).

Craddock, J. M., 1957, "An analysis of the slower temperature variations at Kew Observatory by means of mutually exclusive band pass filters", *Jl R. statist. Soc., Ser. A*, **120**, 387-397.

Craig, J. I., 1912, "The periodogram and method of correlation", *Rep. Br. Ass. Advmt. Sci.*, Report number 82, 416-417.

Craig, J. I., 1916, "A new method of discovering periodicities", *Mon. Not. R. astr. Soc.*, **76**, 493-499.

Cramér, H., 1946, *Mathematical Methods of Statistics* (Princeton University Press, Princeton).

Cressman, G. P., 1958, "Barotropic divergence and very long atmospheric waves", *Mon. Weath. Rev. U.S. Dep. Agric.*, **86**, 293-297.

Cunnyngham, J., 1963, "The spectral analysis of economic time series", US Department of Commerce, Bureau of the Census, Working Paper number 14.

Cutrona, L. J., 1966, "The role of coherent optical systems in data processing", *American Federation of Information Processing Societies, Proceedings of the Spring Joint Computer Conference*, **28**, 24-41.

●Cutrona, L. J., Leith, E. N., Palermo, C. J., Porcello, L. J., 1960, "Optical data processing and filtering systems", *I.R.E. Trans. Inf. Theory*, **II**, 391-400.

Danes, Z. F., Oncley, L. A., 1962, "An analysis of some second derivative methods", *Geophysics*, **27**, 611-615.

Daniell, P. J., 1946, "Symposium on autocorrelation in time series (discussion)", *Jl R. statist. Soc., Ser. B*, **8**, 88-90.

Danielson, G. E., Lanczos, C., 1942, "Some improvements in practical Fourier analysis and their application to X-ray scattering from liquids", *J. Franklin Inst.*, **233**, 365-380, 435-452.

●Darby, E. K., Davies, E. B., 1967, "The analysis and design of two-dimensional filters for two-dimensional data", *Geophys. Prospect.*, **15**, 383-406.

●Darwin, G. H., 1883, "Report of a committee, consisting of Professor G. H. Darwin and J. C. Adams, for the harmonic analysis of tidal observations", *Rep. Br. Ass. Advmt. Sci.*, 49-117.

Davis, H. F., 1963, *Fourier Series and Orthogonal Functions* (Allyn and Bacon, Boston, Mass.).

Davis, H. T., 1941, *The Analysis of Economic Time Series* (Principia Press, Bloomington, Ind.).

Dean, W. C., 1958, "Frequency analysis for gravity and magnetic interpretation", *Geophysics*, **23**, 97-127.

Deland, R. J., 1964, "Travelling planetary waves", *Tellus*, **16**, 271-273.

Derbyshire, J., 1955, "An investigation of storm waves in the North Atlantic Ocean", *Proc. R. Soc., Ser. A*, **230**, 560-569.

Derbyshire, J., 1956, "An investigation into the generation of waves when the fetch of the wind is less than 100 miles", *Q. Jl R. met. Soc.*, **82**, 461-468.

DeVelis, J. B., Reynolds, G. O., 1967, *Theory and Applications of Holography* (Adison-Wesley, Reading, Mass.).

●Dobrin, M. B., Ingalls, A. L., Long, J. A., 1965, "Velocity and frequency filtering of seismic data using laser light", *Geophysics*, **30**, 1144-1178.

●Donohue, J., 1969, "Fourier analysis and the structure of DNA", *Science N.Y.*, **165**, 1091-1096.

Doyle, W., 1962, "Operations useful for similarity-invariant pattern recognition", *J. Ass. comput. Mach.*, **9**, 259-267.

Durbin, J., Watson, G. S., 1951, "Testing for serial correlation in least squares regression II", *Biometrika*, **38**, 159-178.

Elias, P., Grey, D. S., Robinson, D. Z., 1952, "Fourier treatment of optical processes", *J. opt. Soc. Am.*, **42**, 127-134.

Eliasen, E., 1958, "A study of the long atmospheric waves on the basis of zonal harmonic analysis", *Tellus*, **10**, 206-215.

Eliasen, E., Machenhauer, B., 1965, "A study of the fluctuations of the atmospheric planetary flow patterns represented by spherical harmonics", *Tellus*, **17**, 220-238.

Elkins, T. A., 1951, "The second derivative method of gravity interpretation", *Geophysics*, **16**, 29-50.

Estoque, M. A., 1955, "The spectrum of large scale turbulent transfer of momentum and heat", *Tellus*, **7**, 177-185.

●Everett, J. D., 1860, "On a method of reducing observations of underground temperature, with its application to the monthly mean temperatures of underground thermometers, at the Royal Edinburgh Observatory", *Trans. R. Soc., Edinb.*, **22**, 429-439.

Ewing, J. A., 1969, "A note on wavelength and period in confined seas", *J. geophys. Res.*, **74**, 1406-1408.

●Fisher, R. A., 1929, "Tests of significance in harmonic analysis", *Proc. R. Soc., Ser. A*, **125**, 54-59. Also as Paper 16, in *Contributions to Mathematical Statistics* (Wiley, New York).

●Fitzpatrick, E., 1964, "Seasonal distribution of rainfall in Australia analysed by Fourier methods", *Arch. Met. Geophys. Bioklim., Ser. B*, **13**, 270-286.

Flanagan, J. L., 1967, "Spectrum analysis in speech coding", *I.E.E.E. Trans. Audio Electroacoust.*, AU-15, number 2, 66-69.

Fomin, L. M., 1968, "The spectral density of the distribution of drift current velocity in the ocean", *Izvestiya* (Translated in *J. Am. geophys. Un.*), **4**, 735-738.

●Forbes, J. D., 1846, "Account of some experiments on the temperature of the earth at different depths and in different soils near Edinburgh", *Trans. R. Soc., Edinb.*, **16**, 189-236.

●Fourier, J. B. J., 1822, *Theorie analytique de la Chaleur* (see Freeman, A., 1872).

Frankiel, F. N., Schwartzchild, M., 1952, "Preliminary analysis of the turbulence spectrum of the solar photosphere at long wave lengths", *Astrophys. J.*, **116**, 422-427.

Frankiel, F. N., Schwartzchild, M., 1955, *Astrophys. J.*, **121**, 216-223.

Freeman, A. (Transl.), 1872, *The Analytical Theory of Heat* by J. B. J. Fourier (1955 edition, Dover Publications, New York).

●Gentleman, W. M., Sande, G., 1966, "Fast Fourier transforms—for fun and profit", *American Federation of Information Processing Societies Proceedings of the 1966 Fall Computer Conference*, 563-578.

Gifford, F., 1955, "A simultaneous Lagrangian-Eulerian turbulence experiment", *Mon. Weath. Rev. U.S. Dep. Agric.*, **83**, 293-301.

Gilchrist, A., 1957, "The representation of circumpolar 500 mb charts by a series of spherical harmonics", *Brit. Meteorol. Office Meteorol. Res. Papers*, 1040.

Gilman, D. L., Fuglister, F. J., Mitchell, J. M., 1963, "On the power spectrum of 'red noise'", *J. atmos. Sci.*, **20**, 182-184.

Godfrey, M. D., 1965, "An exploratory study of the bi-spectrum of economic time series", *Appl. Statist.*, **14**, 48-69.

Godfrey, M. D., 1967, "Prediction for non-stationary stochastic processes", in *Advanced Seminar on Spectral Analysis of Time Series*, Ed. B. Harris (Wiley, New York), pp.259-269.

Godson, W. L., 1959, "The application of Fourier analysis to meteorlogical data", Tech. Circular number CIR-3168 TEC-295, Met. Branch, Dept. Transport, Canada.

Goertzel, G., 1958, "An algorithm for the evaluation of finite trigonometric series", *Am. math. Mon.*, **65**, 34-35.

●Goldstein, A., Rosenfeld, A., 1964, "Optical correlation for terrain type discrimination", *Photogramm. Engng.*, **30**, 639-646.

Good, I. J., 1958, "The interaction algorithm and practical Fourier series", *Jl R. statist. Soc., Ser. B*, **20**, 361-372.

Good, I. J., 1960, *Jl R. statist. Soc., Ser. B*, **22**, 372-375.

●Goodman, J. W., 1968, *Introduction to Fourier Optics* (McGraw-Hill, New York).

Goodman, N. R., 1957, "On the joint estimation of the spectra, cospectrum and quadrature spectrum of a two-dimensional stationary gaussian process", *Engineering Statistics Laboratory, New York University*, Scientific paper number 10.

Goodman, N. R., 1960, "Measuring amplitude and phase", *J. Franklin Inst.*, **270**, 437-450.

•Goodman, N. R., 1961, "Some comments on spectral analysis of time series", *Technometrics*, **3**, 221-228.

Goodman, N. R., 1963, "Spectral analysis of multiple stationary time series", in *Proceedings of the Symposium on Time Series Analysis*, Ed. M. Rosenblatt (Wiley, New York), pp.260-266.

Goodman, N. R., Katz, S., Kramer, B. H., Kao, M. T., 1961, "Frequency response from stationary noise: two case histories", *Technometrics*, **3**, 245-268.

•Granger, C. W., 1966, "The typical spectral shape of an economic variable", *Econometrica*, **34**, 150-161.

•Granger, C. W., Hatanaka, M., 1964, *An Analysis of Economic Time Series* (Princeton University Press, Princeton).

Granger, C. W., Morgenstern, O., 1963, "Spectral analysis of stock market prices", *Kyklos*, **16**, 1-27.

Grant, F., 1957, "A problem in the analysis of geophysical data", *Geophysics*, **22**, 309-344.

Grenander, U., 1951, "On empirical spectral analysis of stochastic processes", *Ark. Mat.*, **1**, 503-531.

Grenander, U., 1958, "Bandwidth and variance in the estimation of the spectrum", *Jl R. statist. Soc., Ser. B*, **20**, 152-157.

Grenander, U., Rosenblatt, M., 1953, "Statistical spectral analysis of time series arising from stationary stochastic processes", *Ann. math. Statist.*, **24**, 537-558.

Grenander, U., Rosenblatt, M., 1957, *Statistical Analysis of Stationary Time Series* (Wiley, New York).

Griffith, J. L., Panofsky, H. A., Van der Hoven, I., 1956, "Power spectrum analysis over large ranges of frequency", *J. Met.*, **13**, 279-282.

Groves, G., 1955, "Numerical filters for discrimination against tidal periodicities", *Trans. Am. geophys. Un.*, **36**, 1073-1084.

Hamermesh, D. S., 1969, "Spectral analysis of the relation between gross employment changes and output changes, 1958-1966", *Rev. Econ. Statist.*, **51**, 62-69.

•Hamming, R. W., 1962, *Numerical Methods for Scientists and Engineers* (McGraw-Hill, New York).

Hamon, B. V., Hannan, E. J., 1963, "Estimating relations between time series", *J. geophys. Res.*, **68**, 6033-6041.

Hannan, E. J., 1958, "The estimation of the spectrum after trend removal", *Jl R. statist. Soc., Ser. B*, **20**, 323-333.

Hannan, E. J., 1960, *Time Series Analysis* (Methuen, London).

Hannan, E. J., 1961, "Testing for a jump in the spectral function", *Jl R. statist. Soc., Ser. B*, **23**, 394-404.

•Harbaugh, J. W., Preston, F. W., 1965, "Fourier series analysis in geology", *Short Course and Symposium on Computers and Computer Applications in Mining and Exploration*, College of Mines, University of Arizona.

•Hare, F. K., 1960, "The westerlies", *Geogrl. Rev.*, **50**, 345-367.

Harkness, J. P., 1968, "A spectral-analytic test of the long swing hypothesis in Canada", *Rev. Econ. Statist.*, **50**, 429-436.

Harkness, J. P., 1969, "Long swings", *Rev. Econ. Statist.*, **51**, 94-96.

Harper, B. P., 1961, "Energy spectra of 500 mb meridional circulation indices", *J. Met.*, **18**, 487-493.

Harris, B. (Ed.), 1967a, *Advanced Seminar on Spectral Analysis of Time Series* (Wiley, New York).

Harris, B., 1967b, "Introduction to the theory of spectral analysis of time series", *ibid.*, pp.3-23.

●Hartley, H. O., 1949, "Tests of significance in harmonic analysis", *Biometrika*, **36**, 194-201.

●Hasselman, K., Munk, W., Macdonald, G., 1963, "Bi-spectrum of ocean waves", in *Proceedings of the Symposium on Time Series Analysis*, Ed. M. Rosenblatt (Wiley, New York), pp.125-139.

Haubrich, R. A., 1965, "Earth noise, 5 to 500 millicycles per second. 1. Spectral stationarity, normality and nonlinearity", *J. geophys. Res.*, **70**, 1415-1427.

Haubrich, R. A., MacKenzie, G. S., 1965, "Earth noise, 5 to 500 millicycles per second. 2. Reaction of the earth to oceans and atmosphere", *J. geophys. Res.*, **70**, 1429-1440.

Hawkins, J. K., Munsay, C. J., 1963, "Automatic photo reading", *Photogramm. Engng.*, **28**, 632-664.

Helms, H. D., 1967, "Fast Fourier transform method of computing difference equations and simulating filters", *I.E.E.E. Trans. Audio Electroacoust.*, AU-15, number 2, 85-90.

●Henderson, R. G., Zietz, I., 1949, "The computation of second vertical derivatives of geomagnetic fields", *Geophysics*, **14**, 508-516.

Herron, T. J., 1967, "An average geomagnetic power spectrum for period range 4·5 to 12900 seconds", *J. geophys. Res.*, **72**, 759-761.

Hinich, M. J., Clay, C. S., 1968, "The application of the discrete Fourier transform to the estimation of power spectra, coherence and bispectra of geophysical data", *Rev. Geophys.*, **6**, 347-363.

Hogben, N., Lumb, F. E., 1967, *Ocean Wave Statistics* (HMSO, London).

●Holloway, J. L., 1958, "Smoothing and filtering of time and space series", *Adv. Geophys.*, **4**, 351-389.

●Horn, L. H., Bryson, R. A., 1960, "Harmonic analysis of the annual march of precipitation", *Ann. Ass. Am. Geogr.*, **50**, 157-171.

Horwitz, L. P., Shelton, G. L., 1961, "Pattern recognition using autocorrelation", *Proc. Inst. Radio Engrs.*, **49**, 175-185.

Hotelling, H., 1936, "Relations between two sets of variates", *Biometrika*, **28**, 321-377.

Houbolt, J. C., 1961, "Runway roughness studies in the aeronautical field", *J. Air Transp. Div. Am. Soc. Civ. Engrs.*, **87**, 11-31.

●I.E.E.E., 1967, "Special issue on fast Fourier transform and its application to digital filtering and spectral analysis", *I.E.E.E. Trans. Audio Electroacoust.*, AU-15, number 2, 43-114.

Isserlis, L., 1918, "On a formula for the product-moment coefficient of any order of normal frequency distribution in any number of variables", *Biometrika*, **12**, 134-139.

●Jenkins, G. M., 1961, "General considerations in the analysis of spectra", *Technometrics*, **3**, 133-166.

●Jenkins, G. M., 1962, "Cross spectral analysis and the estimation of linear open loop transfer functions", in *Proceedings of the Symposium on Time Series Analysis, Brown University* (Wiley, New York), chapter 18, pp.267-276.

●Jenkins, G. M., 1963, "An example of the estimation of a linear open loop transfer function", *Technometrics*, **5**, 227-245.

●Jenkins, G. M., 1965, "A survey of spectral analysis", *Appl. Statist.*, **14**, 2-32.

Jenkins, G. M. Parzen, E., 1961, "Comments on the discussions of Messrs. Tukey and Goodman", *Technometrics*, **3**, 229-234.

Jenkins, G. M., Priestley, M. B., 1957, "The spectral analysis of time series", *Jl R. statist. Soc.*, Ser. B, **19**, 1-12.

•Jenkins, G. M., Watts, D. G., 1968, *Spectral Analysis* (Holden-Day, San Francisco).

•Jennison, R. G., 1961, *Fourier Transforms and Convolutions for the Experimentalist* (Pergamon Press, Oxford).

•Jones, R. H., 1965, "A reappraisal of the periodogram in spectral analysis", *Technometrics*, 7, 531-542.

Jowett, G. H., 1955, "Sampling properties of local statistics in stationary stochastic series", *Biometrika*, 42, 160-169.

Julian, P. R., 1966, "The index cycle: a cross spectral analysis of zonal index data", *Mon. Weath. Rev. U.S. Dep. Agric.*, 94, 283-293.

Justus, C. G., 1967, "The spectrum and scales of upper atmospheric turbulence", *J. geophys. Res.*, 72, 1933-1940.

Kahn, A. B., 1957, "A generalization of average-correlation methods of spectrum analysis", *J. Met.*, 14, 9-17.

Katz, Y. H., 1964, "Pattern recognition of meteorological satellite cloud photography", *Proceedings of the Third Symposium on Remote Sensing of Environment* (University of Michigan, Ann Arbor), pp.173-214.

•Katz, Y. H., Doyle, W. L., 1964, "Automatic pattern recognition of meteorological satellite cloud photography", Rand Report Memorandum, number RM-3412-NASA.

•Kendall, M. G., 1945, *The Advanced Theory of Statistics* (Charles Griffin, London).

•Kendall, M. G., 1946, *Contributions to the Study of Oscillatory Time Series* (Cambridge University Press, Cambridge).

King, L. J., 1969, "The analysis of spatial form and its relation to geographic theory", *Ann. Ass. Am. Geogr.*, 59, 573-595.

•Kovásznay, L. S. G., Arman, A., 1957, "Optical autocorrelation measurements of two-dimensional random patterns", *Rev. scient. Instrum.*, 28, 793-797.

Krumbein, W. C., 1959, "Trend surface analysis of contour type maps with irregular control-point spacing", *J. geophys. Res.*, 64, 823-834.

Krumbein, W. C., 1963, "Confidence intervals on low-order polynomial trend surfaces", *J. geophys. Res.*, 68, 5869-5878.

•Krumbein, W. C., Graybill, F. A., 1965, *An Introduction to Statistical Models in Geology* (McGraw-Hill, New York).

Kudymov, B. Y., Sokolov, V. A., 1962, *Spectral Well Logging* (Elsevier, New York).

Kung, E. C., Soong, S., 1969, "Seasonal variation of kinetic energy in the atmosphere", *Q. Jl R. met. Soc.*, 95, 501-512.

Lagrange, J. L., 1772a, "Recherches sur la manière de former des table des planètes", *Oeuvres*, 6, 505-627.

Lagrange, J. L., 1772b, "Sur les interpolations", *Oeuvres*, 7, 535-553.

Lahey, J. G., Bryson, R. A., Corzine, H. A., Hutchins, C. W., 1958, *Atlas of 500 mb Wind Characteristics for the Northern Hemisphere* (University of Wisconsin Press, Madison).

•Lanczos, C., 1956, *Applied Analysis*, chapter IV (Prentice-Hall, Englewood Cliffs), pp.207-304.

Lanczos, C., 1966, *Discourse on Fourier Series* (Oliver and Boyd, London).

Landsberg, H. E., Mitchell, J. M., Crutcher, H. L., 1959, "Power spectrum analysis of climatological data for Woodstock College, Maryland", *Mon. Weath. Rev. U.S. Dep. Agric.*, 87, 283-298.

Lederman, W., Reuter, G. E. H., 1954, "Spectral theory for the differential equations of simple birth and death processes", *Phil. Trans. R. Soc., Ser. A*, 246, 321-369.

Lee, W. H. K., Kaula, W. M., 1967, "A spherical harmonic analysis of the earth's topography", *J. geophys. Res.*, 72, 753-758.

•Leese, J. A., Epstein, E. S., 1963, "Application of two-dimensional spectral analysis to the quantification of satellite cloud photographs", *J. appl. Met.*, 2, 629-644.

Lomnicki, Z. A., Zaremba, S. K., 1957, "On estimating the spectral density function of a stochastic process", *Jl R. statist. Soc., Ser. B*, **19**, 13–37.

Lomnicki, Z. A., Zaremba, S. K., 1959, "Bandwidth and resolvability in statistical spectral analysis", *Jl R. statist. Soc., Ser. B*, **21**, 169–171.

Longuet-Higgins, M. S., 1955, "Bounds for the integral of a non-negative function in terms of its Fourier coefficients", *Proc. Camb. phil. Soc. math. phys. Sci.*, **51**, 590–603.

●Longuet-Higgins, M. S., 1957, "The statistical analysis of a random moving surface", *Phil. Trans. R. Soc., Ser. A*, **249**, 321–387.

●Lumley, J. L., Panofsky, H. A., 1964, *The Structure of Atmospheric Turbulence* (Wiley, New York).

MacDonald, N. J., Ward, F., 1963, "The prediction of geomagnetic disturbance indices. I. The elimination of internally predictable variations", *J. geophys. Res.*, **68**, 3351–3373.

McLachlan, D., 1962, "The role of optics in applying correlation functions to pattern recognition", *J. opt. Soc. Am.*, **52**, 454–459.

Maling, G. C., Morney, W. T., Land, W. W., 1967, "Digital determination of third-octave octave and full-octave spectra of acoustical noise", *I.E.E.E. Trans. Audio Electroacoust.*, AU-15, number 2, 98–104.

Matern, B., 1960, "Spatial variation: stochastic models and their application to some problems in forest surveys and other sampling investigations", *Meddn St. SkogsforskInst.*, **49**, number 5.

Mesko, C. A., 1966, "Two-dimensional filtering and the second derivative method", *Geophysics*, **31**, 606–617.

Monin, A. S., 1969, "Ocean turbulence", *Izvestiya* (Translated in *J. Am. geophys. Un.*), **5**, 120–121.

Montgomery, W. D., Brome, P. W., 1962, "Spatial filtering", *J. opt. Soc. Am.*, **52**, 1259–1274.

Munk, W. H., Bullard, E. C., 1963, "Patching the long wave spectrum across tides", *J. geophys. Res.*, **68**, 3627–3634.

●Munk, W. H., Snodgrass, F. E., 1957, "Measurements of southern swell at Guadalupe Is", *Deep Sea Res.*, **4**, 272–286.

Munk, W. H., Snodgrass, F. E., Tucker, M. J., 1959, "Spectra of low frequency ocean waves", *Bull. Scripps Instn. Oceanogr.*, **7**, 283–361.

●N.A.S. (National Academy of Science), 1963, *Ocean Wave Spectra* (Prentice-Hall, Englewood Cliffs).

Neidell, N. S., 1965, "A geophysical application of spectral analysis", *Appl. Statist.*, **14**, 75–88.

Neidell, N. S., 1966, "Spectral studies of marine geophysical profiles", *Geophysics*, **31**, 122–134.

●Neidell, N. S., 1967, "Frequency analysis for sparse and badly sampled data in the earth sciences", *Univ. Kansas State Geol. Surv., Comput. Contr.*, **18**, 15–17.

Nerlove, M., 1964, "Spectral analysis of seasonal adjustment procedures", *Econometrica*, **32**, 241–286.

●Noll, A. M., 1964, "Short-time spectrum and 'cepstrum' techniques for vocal-pitch detection", *J. acoust. Soc. Am.*, **36**, 296–302.

Nyquist, H., 1928, "Certain topics in telegraph transmission theory", *Trans. Am. Inst. elect. Engrs.*, **47**, 617–644.

●O'Neill, E. L., 1956, "Spatial filtering in optics", *I.R.E. Trans. Inf. Theory*, **2**, 56–65.

O'Neill, E. L., 1963, *Introduction to Statistical Optics* (Adison Wesley, Reading, Mass.).

Oort, A. H., Taylor, A., 1969, "On the kinetic energy spectrum near the ground", *Mon. Weath. Rev. U.S. Dep. Agric.*, **97**, 623–636.

Page, A. N., 1967, "Residential construction: exploration of the statistical series", *J. Business*, **40**, 36-43.

Page, C. G., 1952, "Instantaneous power spectra", *J. appl. Phys.*, **23**, 103-106.

●Panofsky, H. A., 1953, "The variation of the turbulence spectrum with height under superadiabatic conditions", *Q. Jl R. met. Soc.*, **79**, 150-153.

Panofsky, H. A., 1955, "Meteorological applications of power-spectrum analysis", *Bull. Am. met. Soc.*, **36**, 163-166.

Panofsky, H. A., 1967, "Meteorological applications of cross-spectrum analysis", in *Advanced Seminar on Spectral Analysis of Time Series*, Ed. B. Harris (Wiley, New York), pp.109-132.

Panofsky, H. A., Brier, G. W., 1958, *Some Applications of Statistics to Meteorology* (Pennsylvania State University Press, Penn.).

Panofsky, H. A., Deland, R. J., 1959, "One-dimensional spectra of atmospheric turbulence in the lowest 100 meters", *Adv. Geophys.*, **6**, 41-64.

●Panofsky, H. A., McCormick, R. A., 1954, "Properties of spectra of atmospheric turbulence at 100 metres", *Q. Jl R. met. Soc.*, **80**, 546-564.

Panofsky, H. A., Wolff, P., 1957, "Spectrum and cross spectrum analysis of hemispheric westerly index", *Tellus*, **9**, 195-200.

Parzen, E., 1956, "On consistent estimates of the spectral density of a stationary time series", *Proc. natn. Acad. Sci. U.S.A.*, **42**, 154-157.

Parzen, E., 1957a, "On choosing an estimate of the spectral density function of a stationary time series", *Ann. math. Statist.*, **28**, 921-932.

Parzen, E., 1957b, "On consistent estimates of the spectrum of a stationary time series", *Ann. math. Statist.*, **28**, 329-348.

Parzen, E., 1957c, "A central limit theorem for multilinear stochastic processes", *Ann. math. Statist.*, **28**, 252-255.

Parzen, E., 1957d, "Optimum consistent estimates of the spectrum of a stationary time series", *Ann. math. Statist.*, **28**, 329-348.

Parzen, E., 1958, "On asymptotically efficient consistent estimates of the spectral density function of a stationary time series", *Jl R. statist. Soc., Ser. B*, **20**, 303-322.

Parzen, E., 1961a, "Mathematical considerations in the estimation of spectra", *Technometrics*, **3**, 167-190.

Parzen, E., 1961b, "An approach to time series analysis", *Ann. math. Statist.*, **32**, 951-989.

Parzen, E., 1967a, "Time series analysis for models of signal plus white noise", in *Advanced Seminar on Spectral Analysis of Time Series*, Ed. B. Harris (Wiley, New York), pp.233-257.

Parzen, E., 1967b, *Time Series Analysis Papers* (Holden-Day, San Francisco).

●Peixoto, J. P., Saltzman, B., Teweles, S., 1964, "Harmonic analysis of the topography along parallels of the earth", *J. geophys. Res.*, **69**, 1501-1505.

Phillips, O. M., 1958, "The equilibrium range in the spectrum of wind generated waves", *J. Fluid. Mech.*, **4**, 426-434.

●Pierson, W. J. (Ed.), 1960, "The directional spectrum of a wind generated sea as determined from data obtained by the stereo wave observation project", *Met. Pap. N.Y. Univ.*, **2**, 6.

Pierson, W. J., Marks, W., 1952, "The power spectrum analysis of ocean wave records", *Trans. Am. geophys. Un.*, **33**, 834-844.

Pierson, W. J., Moskowitz, L., 1964, "A proposed spectral form for fully developed wind seas based on the similarity theory of S. A. Kitaigorodskii", *J. geophys. Res.*, **69**, 5181-5190.

●Pincus, H., Dobrin, M. B., 1966, "Geological applications of optical data processing", *J. geophys. Res.*, **70**, 4861-4869.

Platzman, G. W., 1960, "The spectral form of the vorticity equation", *J. Met.*, 17, 635-644.

Platzman, G. W., Rao, D. B., 1964, "Spectra of Lake Erie water levels", *J. geophys. Res.*, 69, 2525-2535.

Pollock, D.K., Koestler, C.J., Tippett, J.T., 1963, *Optical Processing of Information* (Spartan Books, Baltimore).

Polowchak, van M., Panofsky, H. A., 1968, "The spectrum of daily temperatures as a climatic indicator", *Mon. Weath. Rev. U.S. Dep. Agric.*, 96, 596-600.

Press, H., Houbolt, J. C., 1955, "Some applications of generalized harmonic analysis to gust loads on airplanes", *J. aeronaut. Sci.*, 22, 17-26.

●Press, H., Tukey, J. W., 1956, "Power spectral methods of analysis and their application to problems in airplane dynamics", in *Flight Test Manual*, Ed. E. J. Durbin, NATO, Advisory Group for Aeronautics Res. and Development. Also Bell System Monograph, number 2606.

●Preston, F. W., 1966, "Two-dimensional power spectra for classification of land forms", *Univ. Kansas State Geol. Surv., Comput. Contr.*, 7, 64-69.

Preston, F. W., Harbaugh, J. W., 1965, "BALGOL programs and geologic application for single and double Fourier series using IBM 7090/7094 computers", *Univ. Kansas State Geol. Surv., Comput. Contr.*, Special Publication number 24.

Priestley, C. H. B., 1958, *Turbulent Diffusion in the Atmosphere* (Chicago University Press, Chicago).

Priestley, M. B., 1962, "The analysis of stationary processes with mixed spectra I-II", *Jl R. statist. Soc., Ser. B*, 24, 215-233, 511-529.

Priestley, M. B., 1962, "Basic considerations in the estimation of spectra", *Technometrics*, 4, 551-564.

Priestley, M. B., 1965, "The role of bandwidth in spectral analysis", *Appl. Statist.*, 14, 33-47.

Quandt, R. E., 1964, "Statistical discrimination among alternative hypotheses and some economic regularities", *J. reg. Sci.*, 5, 1-24.

Quimpo, R. G., 1968a, "Stochastic analysis of daily river flows", *J. Hydraul. Div. Am. Soc. civ. Engrs.*, 94, number HY1, paper 5719.

Quimpo, R. G., 1968b, "Autocorrelation and spectral analysis in hydrology", *J. Hydraul. Div. Am. Soc. civ. Engrs.*, 94, number HY2, paper 5837.

●Rayleigh, Lord, 1880, "On the resultant of a large number of vibrations of the same pitch and arbitrary phase", *Phil. Mag.*, 10, 73-78.

●Rayner, J. N., 1965, *An Approach to the Dynamic Climatology of New Zealand*, University of Canterbury, New Zealand (unpublished).

Rayner, J. N., 1967a, "Correlation between surfaces by spectral methods", *Univ. Kansas State Geol. Surv., Comput. Contr.*, 12, 31-37.

●Rayner, J. N., 1967b, "A statistical model for the explanatory description of large scale time and spatial climate", *Can. Geogr.*, 11, 68-86.

Rhodes, J., 1953, "Analysis and synthesis of optical images", *Am. J. Phys.*, 21, 337-343.

Rice, S. O., 1944, "Mathematical analysis of random noise", *Bell Syst. tech. J.*, 23, 24; also published in *Selected Papers on Noise and Stochastic Process*, Ed. N. Wax (Dover Publications, New York).

Rodriguez-Iturbe, I., Nordin, C. F., 1968, "Time series analysis of water and sediment discharges", *Bull. int. Ass. scient. Hydrol.*, 31, 2.

Rodriguez-Iturbe, I., Yovjevich, V., 1968, "The investigation of relationship between hydrologic time series and sunspot numbers", *Colorado State Univ., Hydrol. Papers*, number 26.

Roetling, P. G., Hemmil, H. B., 1962, "Study of spatial filtering by optical diffraction for pattern recognition", *Cornell Aeronautical Laboratory Report.*

Rogerson, J. B., 1955, "A statistical analysis of spectroheliograms", *Astrophys. J.,* **121**, 204-215.

Rosenback, O., 1953, "A contribution to the computation of 'Second Derivative' from gravity data", *Geophysics,* **18**, 894-912.

Rosenblatt, M., 1959, "Statistical analysis of stochastic processes with stationary residuals", in *Probability and Statistics,* Ed. U. Grenander (Wiley, New York).

Rosenblatt, M., 1963, *Proceedings of the Symposium on Time Series Analysis* (Wiley, New York).

Rosenblatt, M., 1965, "Spectral analysis and parametric methods for seasonal adjustment of economic time series", US Department of Commerce, Bureau of the Census, Working Paper number 23.

Rosenblatt, M., 1966, "Remarks on higher order spectra", in *Multivariate Analysis,* Ed. P. R. Krishnaiah (Academic Press, New York), pp.383-388.

Rosenfeld, A., 1962, "Automatic recognition of basic terrain types from aerial photographs", *Photogramm. Engng.,* **27**, 115-132.

Rosenfeld, A., 1965, "Image processing", in *Proceedings of the Third Annual Conference on Urban Planning Information Systems and Programs* (The American Society of Planning Officials, Chicago).

Rosenfeld, A., Fried, C., Orton, N., 1965, "An approach to the automatic interpretation of cloud cover pictures", *Photogramm. Engng.,* **31**, 991-1002.

Rosenfeld, A., Goldstein, A., 1964, "Optical correlation for terrain type discrimination", *Photogramm. Engng.,* **30**, 639-646.

Rosenfeld, A., Pfaltz, J. L., 1966, "Sequential operations in digital picture processing", *J. Ass. comput. Mach.,* **13**, 471-494.

Royal Statistical Society, 1965, "Special issue on spectral analysis", *Appl. Statist.,* **14**.

Rudwick, P., 1966, "Note on the calculation of Fourier series", *Maths. Comput.,* **20**, 429-430.

●Runge, C., 1903, "Über die Zerlegung empirisch gegebener periodischer Funktionen in Sinuswellen", *Z. Math. Phys.,* **48**, 443-456.

Rushton, S., Neumann, J., 1957, "Some applications of time series analysis to atmospheric turbulence and oceanography", *Jl R. statist. Soc., Ser. A,* **120**, 409-425.

●Sabbagh, M., Bryson, R., 1962, "Aspects of the precipitation climatology of Canada investigated by the method of harmonic analysis", *Ann. Ass. Am. Geogr.,* **52**, 426-440.

Saltzman, B., 1957, "Equations governing the energetics of the larger scales of atmospheric turbulence in the domain of wave number", *J. Met.,* **14**, 513-523.

Saltzman, B., 1958, "Some hemispheric spectral statistics", *J. Met.,* **15**, 259-263.

Saltzman, B., Fleisher, A., 1960, "Spectrum of kinetic energy transfer due to large scale horizontal Reynolds stresses", *Tellus,* **12**, 110-111.

Saltzman, B., Teweles, S., 1964, "Further statistics on the exchange of kinetic energy between harmonic components of the atmosphere flow", *Tellus,* **16**, 432-435.

●Schuster, A., 1897, "On lunar and solar periodicities of earthquakes", *Proc. R. Soc.,* **61**, 455-465.

●Schuster, A., 1898, "On the investigation of hidden periodicities with application to a supposed 26-day period of meteorological phenomena", *Terr. Magn. atmos. Elect.,* **3**, 13-17.

●Schuster, A., 1900, "The periodogram of magnetic declination", *Trans. Camb. phil. Soc.,* **18**, 107-135.

●Schuster, A., 1906a, "The periodogram and its optical analogy", *Proc. R. Soc., 77*, 136–140.

●Schuster, A., 1906b, "On the periodicities of sunspots", *Phil. Trans. R. Soc., Ser. A, 206*, 69–100.

●Schwerdtfeger, W., Prohaska, F., 1956, "The semi-annual pressure oscillation, its cause and effects", *J. Met., 13*, 217–218.

●Seiwell, H. R., 1949, "The principles of time series analysis applied to ocean wave data", *Proc. natn. Acad. Sci. U.S.A., 35*, 518–528.

Seiwell, H. R., Wadsworth, G. P., 1949, "A new development in ocean wave research", *Science, N.Y., 109*, 271–274.

Scott, J. R., 1969, "Some average wave lengths on short crested seas", *Q. Jl R. met. Soc., 95*, 621–634.

●Shapiro, J. M., Whitney, D. R., 1967, *Elementary Analysis and Statistics* (Merrill Books, Columbus, Ohio).

Sharma, B., Geldart, L. P., 1966, "Analysis of gravity anomalies of two dimensional faults using Fourier transforms", *Geophys. Prospect., 16*, 77–93.

Silberman, I., 1954, "Planetary waves in the atmosphere", *J. Met., 11*, 27–34.

●Singleton, R. C., 1967, "A method for computing the fast Fourier transform with auxiliary memory and limited high-speed storage", *I.E.E.E. Trans. Audio Electroacoust.*, AU-15, number 2, 91–98.

Singleton, R. C., Poulter, T. C., 1967, "Spectral analysis of the call of the male killer whale", *I.E.E.E. Trans. Audio Electroacoust.*, AU-15, number 2, 104–113.

Slutzky, E., 1937, "The summation of random causes on the source of cyclic processes" (Translated from Russian), *Econometrica, 5*, 105–146.

Soong, S. T., Kung, E. C., 1969, "Short period kinetic energy cycles in the atmosphere", *J. appl. Met., 8*, 484–491.

Southworth, R. W., 1959, "Autocorrelation and spectral analysis", in *Mathematical Methods for Digital Computers*, Eds. A. Ralston and H. S. Wilf (Wiley, New York), pp.213–220.

●Speight, J. G., 1965, "Meander spectra of the Angabunga river", *J. Hydrol., 3*, 1–15.

Speight, J. G., 1967, "Spectral analysis of meanders of some Australian rivers", in *Landform Studies from Australia and New Guinea*, Eds. J. N. Jennings and J. A. Marbbutt (Australian National University Press), pp.48–63.

Stockham, T. G., 1966, "High-speed convolution and correlation", *American Federation of Information Processing Societies Proceedings of the Spring Joint Computer Conference, 28*, 229–233.

Stone, R., Dugundi, J., 1965, "A study of microrelief in a mapping classification, and quantification by means of a Fourier analysis", *Engng. Geol., 1*, 91–187.

Stroke, G. W., 1963, *An Introduction to Coherent Optics and Holography* (Academic Press, New York).

Strutt, J. W., 1871, "Note on the exploration of coronas as given in Verdet's 'Lecors d'Optique physique', and other works", *Proc. Lond. math. Soc., 4*, 253–283.

●Taylor, G. I., 1920, "Diffusion by continuous movement", *Proc. Lond. math. Soc., Ser. A, 20*, 196–212.

Taylor, G. I., 1935, "Statistical theory of turbulence", *Proc. R. Soc., Ser. A, 151*, 421–478.

●Taylor, G. I., 1938, "The spectrum of turbulence", *Proc. R. Soc., Ser. A, 164*, 476–490.

●Thomson, W., 1860, "On the reduction of observations of underground temperatures; with application to Professor Forbes' Edinburgh observations, and the continued Calton Hill series No. XVII", *Trans. R. Soc., Edinb., 22*, 405–427.

Tick, L. J., 1961, "The estimation of transfer functions of quadratic systems", *Technometrics, 3*, 563.

Tick, L. J., 1962, "Conditional spectra, linear systems and coherency", in *Proceedings of the Symposium on Time Series Analysis,* Ed. M. Rosenblatt (Wiley, New York), pp.197-203.

Tick, L. J., 1967, "Estimation of coherency", in *Advanced Seminar on Spectral Analysis of Time Series,* Ed. B. Harris (Wiley, New York), pp.133-153.

•Tippett, J. T., Berkowitz, D. A., Clapp, L. C., Koester, C. J., Vanderburgh, A., 1965, "Optical and electro-optical information processing", *Proceedings of a Symposium held in Boston in 1964* (MIT Press, Cambridge, Mass.).

Tobler, W. R., 1968, "A digital terrain library", Technical Report, University of Michigan, ORA Project 08055, under contract with US Army Res. Off. (Durham), contract number DA-31-124-ARO-D-456, Durham, North Carolina.

•Tobler, W. R., 1969a, "Geographical filters and their inverses", *Geogr. Anal.,* 1, 234-253.

•Tobler, W. R., 1969b, "An analysis of a digitalized surface", in *A Study of Land Type,* Ed. C. M. Davis, Final Report, University of Michigan, ORA Project 08055, under contract with US Army Res. Off. (Durham), contract number DA-31-124-ARO-D-456, Durham, North Carolina.

•Tobler, W. R., 1969c, "The spectrum of US40", *Papers and Proceedings of The Regional Science Association,* 23, 45-52.

•Tolstov, G. P., 1962, *Fourier Series,* translated from Russian by R. A. Silverman (Prentice-Hall, Englewood Cliffs).

•Tukey, J. W., 1949, "The sampling theory of power spectrum estimates", in *Symposium on Applications of Auto Correlation Analysis to Physical Problems,* O.N.R. Woods Hole, NAVEXOS-P-735, 47-67.

Tukey, J. W., 1959a, "An introduction to the measurement of spectra", in *Probability and Statistics,* Ed. U. Grenander (Wiley, New York), pp.300-330.

Tukey, J. W., 1959b, "The estimation of (power) spectra and related quantities", in *On Numerical Approximation,* Ed. R. E. Langer (University of Wisconsin Press, Madison), pp.389-411.

•Tukey, J. W., 1961, "Discussion, emphasizing the connection between analysis of variance and spectrum analysis", *Technometrics,* 3, 191-220.

Tukey, J. W., 1965, "Data analysis and the frontiers of geophysics", *Science, N.Y.,* 148, 1282-1289.

•Tukey, J. W., 1967, "Spectrum calculations in the new world of the fast Fourier transform", in *Advanced Seminar on Spectral Analysis of Time Series,* Ed. B. Harris (Wiley, New York), pp.25-46.

•Uberoi, M. S., 1955, "Investigation of turbulence in the solar atmosphere", *Astrophys. J.,* 121, 440-503; 122, 466-476.

Van der Hoven, I., 1957, "Power spectrum of horizontal wind speed in the frequency range from 0.0007 to 900 cycles per hour", *J. Met.,* 14, 160-164.

Van Isacker, J., 1961, "Generalised harmonic analysis", *Adv. Geophys.,* 7, 189-214.

•Van Loon, H., Jeune, R. L., 1969, "The half-yearly oscillations in the tropics of the southern hemisphere", *J. atmos. Sci.,* 26, 218-232.

•Walker, G., 1914, "On the criteria for the reality of relationships or periodicities", *Calcutta Ind. Met. Memo,* number 21, part 9.

Ward, F. W., 1960, "The variance (power) spectra of C_i, K_p and A_p", *J. geophys. Res.,* 65, 2359-2373.

Ward, F., Shapiro, R., 1961, "Meteorological periodicities", *J. Met.,* 18, 635-656.

•Weatherburn, C. E., 1962, *A First Course in Mathematical Statistics* (Cambridge University Press, Cambridge).

Webb, E. K., 1955, "Autocorrelations and energy spectra of atmospheric turbulence", Div. Met. Physics Tech. Paper number 5, Commonwealth Scientific and Indust. Res. Org. Melbourne.

Welch, P. D., 1961, "A direct digital method of power spectrum estimation", *IBM Jl Res. Dev.,* **5**, 141-156.

Welch, P. D., 1967, "The use of the fast Fourier transform for the estimation of power spectra: a method based on time averaging over short modified periodograms", *I.E.E.E. Trans. Audio Electroacoust.,* AU-15, number 2, 70-73.

Welch, L. R., 1965, "Computation of finite Fourier series", *Jet propul. Lab.,* Report number SPS 37-37, volume IV, 295-297.

Whittaker, E. T., Robinson, G., 1924, *The Calculus of Observations* (Blackie and Sons, London).

Whittle, P., 1957, "Curve and periodogram smoothing", *Jl R. statist. Soc., Ser. B,* **19**, 38-47.

●Wiener, N., 1930, "Generalized harmonic analysis", *Acta. math., Stockh.,* **55**, 117-258.

Wiener, N., 1933, *The Fourier Integral and Certain of Its Applications* (Dover Publications, New York).

Wiener, N., 1949, *Time Series* (MIT Press, Cambridge, Mass.).

Wiener, N., 1958, *Nonlinear Problems in Random Theory* (Wiley, New York).

Wiener, N., 1966, in *Generalised Harmonic Analysis and Tauberian Theorems* MIT Paperback number 51 (MIT Press, Cambridge, Mass.).

●Wiin-Nielsen, A., 1959, "A study of energy conversion and meridional circulation for the large-scale motion in the atmosphere", *Mon. Weath. Rev. U.S. Dep. Agric.,* **87**, 319-332.

●Wiin-Nielsen, A., Brown, J. A., Drake, M., 1963, "On atmospheric energy conversions between the zonal flow and the eddies", *Tellus,* **15**, 261-279.

●Wiin-Nielsen, A., Brown, J. A., Drake, M., 1964, "Further studies of energy exchange between the zonal flow and the eddies", *Tellus,* **16**, 168-180.

Winston, J. S., 1960, "Some new data on the longitudinal dimensions of planetary waves", *J. Met.,* **17**, 522-531.

Wold, H., 1954, *A Study in the Analysis of Stationary Time Series* (Almquist and Wiksell, Stockholm).

●Wold, H., 1965, *Bibliography on Time-Series and Stochastic Processes* (MIT Press, Cambridge, Mass.).

Wonnacott, T. A., 1961, "Spectral analysis combining a Bartlett window with an associated inner window", *Technometrics,* **3**, 235-243.

Yen, Y-C, Dotson, J. W., 1969, "Harmonic analysis of snow temperatures by Fourier techniques", *J. geophys. Res.,* **68**, 2217-2232.

Yevdjevich, V. M., Roesner, L. A., 1966, "Mathematical models for time series of monthly precipitation and monthly runoff", *Colorado State Univ., Hydrol. Papers,* number 15.

Yule, G. U., 1897, "On the theory of correlation", *Jl R. statist. Soc.,* **60**, 812-854.

Yule, G. U., 1921, "On the time correlation problem", *Jl R. statist. Soc.,* **84**, 497-537.

Yule, G. U., 1926, "Why do we sometimes get nonsense correlations between time series?", *Jl. R. statist. Soc.,* **89**, 1-64.

Zaremba, S. K., 1967, "Quartic statistics in spectral analysis", in *Advanced Seminar on Spectral Analysis of Time Series,* Ed. B. Harris (Wiley, New York,), pp.47-79.

APPENDIX A

List and definition of symbols

Most symbols are general and apply throughout the text. A few, which are restricted to a page or two, are distinguished by the letter R after the page number(s). Note that in subscripted variables such as $a[k]$ the definition is given for one dimension (one subscript) only, but that it will apply to more dimensions, i.e. $a[k_1, k_2]$. The page number quoted in the second column is where the symbol first appears in the text.

Symbol	Page	Definition
$a[f]$	56	Amplitude density of even function at frequency f. See $a[k]$.
$a[k]$	16	Amplitude of cosine waves at the discrete frequency k, an integer, i.e. cosine amplitude as a function of k, frequency. For two dimensions this becomes $a[k_1, k_2]$. Subscripts refer to the variable which contains the cosine wave, e.g. $a_x[k]$. The letter f, not necessarily an integer, as in $a_x[f]$, refers to the continuous frequency whose centre, from a particular band, lies at f.
$a_x[k]$	52	See $a[k]$.
$a'[k]$	56	The estimate of $a[k]$ after zeros are added to the data.
$\hat{a}[k]$	82	The estimate of $a[k]$ after zeros are added and a window is applied to the data.
$\bar{a}[k]$	17, 23	$= a[k]/2$, where $k = 0$ and $k = n/2$. This notation has been introduced so that *all* $a[k]$ may be calculated from the same equations (3.3.11) or (9.2.6).
A	15R	Amplitude of sinusoidal curve.
A	18R	Amplitude, a vector magnitude.
$A[k]$	16	Amplitude of the combined sine and cosine waves at frequency k. $A[k] = (a^2[k] + b^2[k])^{\frac{1}{2}}$.

Symbol	Page	Definition
$b[f]$	57	Amplitude density of odd function at frequency f. See $b[k]$.
$b[k]$	16	Amplitude of sine wave at frequency k. See $a[k]$.
$\hat{b}[k]$	82	See equivalent a.
B	41R	Amplitude, a vector magnitude.
c_1, c_2, etc.	35R	Arbitrary constants.
$c[f]$	57	Complex amplitude density at frequency f. $c[f] = a[f] - ib[f]$.
$c[k]$	38	Complex amplitude at frequency k. $c[k] = (a[k] - ib[k])/2$.
cis	41	Refers to $\cos \pm i\sin$. The circular spectral function.
D	82	Number of actual observations.
D subscript	69	Used with weighting function to denote equally spaced weights, i.e. a Dirac function.
e	34R	Exponential number $2 \cdot 718 \ldots$
$\exp(t)$	34	e to the power of t, i.e. e^t.
$\mathcal{E}(\ )$	45	$= \exp\{-i2\pi(\ )\}$
$E(\ )$	75R	Expected value of the function in the brackets.
E	88	Prewhitening constant.
E_1, E_2	116	Prewhitening constants in two dimensions
E subscript	42	Denoting an even function.
f	57	Frequency, a continuous variable. The central frequency of a band of continuous frequencies usually in units of cycles per units of the independent variable such as $t, f = k/n\Delta t$ (e.g. let $n = 10$, $k = 2$, and $\Delta t = 5$ seconds, then $f = 0 \cdot 04$ cycles per second). For two dimensions this becomes f_1, f_2.
f_1, f_2	109	f for two dimensions. See f.

Symbol	Page	Definition
f_{12}	104	f for two dimensions along the normal to the set (block) of waves.
f_1	60	Frequency, a continuous variable. Similar to f but independent of it.
$F[\ ]$	25, 99	The F distribution.
F subscript	73	Used with f to denote the folding frequency, the Nyquist frequency, $f_F = 1/2\Delta p$, where $\Delta p = $ constant $\times \Delta t$.
$\pm g$	98	The deviations from the mean for a specified cumulative magnitude (confidence level) of the standard normal distribution.
G	83	Number of tapering points. Both ends of a series will be tapered by this amount.
G_1, G_2	116	Tapering points in two dimensions.
$h[t]$	59	Function of t.
$H[f]$	59	Fourier transform of $h[t]$, i.e. $$H[f] = \int_{-\infty}^{\infty} h[t]\exp(-i2\pi ft)\,dt \ .$$
i	36	Imaginary number $i = \sqrt{-1}$.
I subscript	38R	Referring to the imaginary part of a function.
I	70	Integration interval $= 2M\Delta t$.
j	20, 20	Independent variable, frequency, an integer, which refers to the dependent variable, say x, e.g. $x[j]$ is the $(j+1)$th observation of x. Therefore n equally spaced observations of x may be labelled $x[0], x[1], \dots , x[j], \dots , x[n-1]$ or $$x\left[-\left(\frac{n+1}{2}-1\right)\right], x\left[-\left(\frac{n+1}{2}-2\right)\right], \dots ,$$ $$x[-j], \dots , x[0], \dots , x[j], \dots , x\left[\frac{n+1}{2}-1\right]$$ for n odd. For two dimensions an observation of x is written $x[j_1, j_2]$.

Symbol	Page	Definition
		N.B. the actual independent variable is often not an integer. Therefore a relationship must be set up between the integer independent variable j and the actual independent variable, say t. In other words Δt is made equal to 1 and t for $j = 0$ is made equal to 0. Thus $x[t[j]]$ is equivalent to $x[j]$.
j_1, j_2	102	j for two dimensions. See j.
j_0, j_1, j_2	45	Integer variables which combine to equal j.
J	76R	Variable to distinguish different sequences of data.
k	15	Integer which refers to frequency. Usually in cycles per basic interval. The basic interval is n observations or 2π radians. Frequencies equivalent to two dimensions are k_1, k_2.
k_1, k_2	102	k for two dimensions. See k.
k_{12}	104	Frequency along the normal to two-dimensional waves. $k_{12} = (k_1^2 + k_2^2)^{½}$.
k'	39R	A particular frequency.
k'	56R	A particular frequency $= 2k$. See k.
k_{max}	21	Maximum k calculable $= n/2$ (n even), $= (n-1)/2$ (n odd).
l	44	Length of cycle.
L subscript	73R	Refers to an aliased function, a function dependent upon frequency such that the function at a given calculable frequency is the sum of the function at all corresponding aliased frequencies.
m	79	Maximum lag. In two dimensions this becomes m_1, m_2.
m_1, m_2	117	m for two dimensions. See m.
m subscript	47R	Denotes $(m+1)$th r factor of n.

Symbol	Page	Definition
M	70	A number of equally spaced data points such that $2M\Delta t = I$. Therefore the total number of data points is $2M+1$.
n	20, 102	Number of observations. In two dimensions this becomes n_1, n_2.
n'	55R	Number of observations $n' = 2n$.
N	76R	Number of different sequences of data.
O subscript	42	Denoting an odd function.
p	75	Lag.
P	41R	Point.
q	93	Lag.
r	81	Integer which refers to frequency in the same sense as k but not equal to k. Whereas $k_{max} = n/2$, $r_{max} = m$, or $= (n/2)/(2z+1)$.
r_1, r_2	117	Integers for frequency in two dimensions. See r.
r_0, r_1, etc.	45	Factors of n.
r_{yx}^2	94	Coefficient of determination.
$\hat{R}_{yx}[r]$	94	Coherence, $\hat{R}_{yx}^2[r] = \|\hat{XY}[r]\|^2/(\hat{XX}[r]\,\hat{YY}[r])$.
R subscript	38R	Referring to the real part of a function.
R	16R	An arbitrary angle.
S	16R	An arbitrary angle.
t	20	Independent variable such as distance or time. In two dimensions this becomes t_1, t_2.
t_1, t_2	19, 102	t for two dimensions. See t.
T	19	Total extent of independent variable $= n\Delta t$, i.e. length of basic interval 2π.
T_1, T_2	103	T for two dimensions. See T.
u, v	21R, 34R, 66R, 93R	Arbitrary parameters.

Symbol	Page	Definition
V	12	Vector.
$V[\ ]$	94	Transform of $v[\ ]$.
$\hat{V}_{yx}[\ ]$	96	The effect on y of a change in x, $\hat{V}_{yx}[\ ] = |\hat{X}\hat{Y}[\ ]|/\hat{X}\hat{X}[\ ]$.
$w[t]$	66	Function of t. Usually a weighting function. Subscripts are used to denote different weighting functions.
$W^*[f]$	68	Fourier transform of $w[t]$, $$W^*[f] = \int_{-\infty}^{\infty} w[t]\exp(-\mathrm{i}2\pi ft)\mathrm{d}t.$$
$x[t]$	19	Dependent variable, a function of t.
$x'[t]$	82	Sample of $x[t]$.
$x'[t], x''[t]$	26R	Specific portions of $x[t]$.
$x[\theta]$	17, 21	Dependent variable, function of θ.
$xx[p]$	75	Dependent variable, function of p. Average lagged product, autocovariance, $$xx[p] = \lim_{n \to \infty}\int_{-\infty}^{\infty} x[t]x[t+p]\mathrm{d}t.$$
$xx'[p]$	78	Dependent variable, function of p. Calculable autocovariance, $$xx'[p] = \frac{1}{n-|p|}\sum_{j=0}^{n-1-|p|} x[j]x[j+p],$$ i.e. estimate of $xx[p]$ from a sample.
$\hat{xx}[p]$	79	Dependent variable, function of p. Estimated value of autocovariance, $\hat{xx}[p] = xx'[p]h[p]$.
X	13R	Abscissa in the Cartesian system.
$X[f]$	58	Fourier transform of $x[t]$. $$X[f] = \int_{-\infty}^{\infty} x[t]\exp(-\mathrm{i}2\pi ft)\mathrm{d}t.$$
$XX[f]$	77	Fourier transform of $xx[p]$.
$X'[f]$	82	Fourier transform of the sample $x'[t]$.
$XX'[f]$	79	Fourier transform of $xx'[p]$.

Symbol	Page	Definition
$\hat{X}X[f]$	79	Estimate of the spectrum of $x[t]$. Fourier transform of $xx[p]$. $\hat{X}X[f] = XX'[f] * H[f]$ $= \int_{-\infty}^{\infty} XX[f]H[f_1-f]\,df,$ a better estimate of $XX[f]$ than $XX'[f]$. Note that $\hat{X}X[f]$ is aliased. Also, $\hat{X}X[f] = |\hat{X}[f]|^2$.
$\tilde{X}X[f]$	86	Filtered true spectrum, $\tilde{X}X[f] = XX[f]H[f_1-f]$.
$xy[p]$	94	Lag covariance function, function of p. Average lagged product. $xy[p] = \lim_{n \to \infty} \frac{1}{n}\int_{-\infty}^{\infty} x[t]y[t+p]\,dt.$
$xy'[p]$	95	Calculable lag covariance function.
$xy'_E[p]$	95	Even part of $xy'[p]$.
$xy'_O[p]$	95	Odd part of $xy'[p]$.
$XY[f]$	94	Transform of $xy[p]$ (complex).
$XY'_O[\ ]$	96	Fourier transform of $xy'_O[p]$. Sine transform.
$\hat{X}Y_E[\ ]$	95	Estimate of the cospectrum.
$\hat{X}Y_O[\ ]$	96	Estimate of the quadrature spectrum.
$|\hat{X}Y[\ ]|$	96	Estimate of the cross spectrum, $|\hat{X}Y[\ ]| = (\hat{X}Y_E^2[\ ] + \hat{X}Y_O^2[\ ])^{\frac{1}{2}}$. Compare with $A[k]$.
y		may replace x in all definitions, which are general and which involve only x above. Denotes another dependent variable.
$y'[t], y''[t]$, etc.	35R	Differentiated values of $y[t]$, i.e. $y''[t] = \frac{d(y'[t])}{dt} = \frac{d^2}{dt^2}(y[t]).$
Y		may replace X in all definitions, which are general and which involve only X above. Denotes transform of another variable.
Y	12R	Ordinate in the Cartesian system.

Symbol	Page	Definition
z	83	A number used in averaging.
$z1, z2$	117	z for two dimensions. See z.
α	40	Coordinate in a complex plane.
β	40	Coordinate in a complex plane.
$\Gamma(\ )$	84	The gamma function $\Gamma(n) = (n-1)!$
δ	69	Infinitely small change in.
Δ	20	Increment in.
θ	12R, 20R	Angle independent variable.
Θ	104	Orientation from the abscissa of the normal to two-dimensional waves.
μ	75	Population mean.
ν	84	Degrees of freedom.
π	12	$3 \cdot 14159 \ldots$ radians.
$\hat{\sigma}^2$	24	Estimated variance.
Σ	17	Summation.
$\tau[\ ]$	96	Lag between series, distance between maxima in x and in y at a given frequency, $\tau[r] = \Phi[f]/2\pi f = (\Phi[r]/2\pi r)(2m\Delta t).$
$\Phi[\ ]$	15, 96	Phase angle. Distance from the origin to the first crest, assuming that for any frequency one cycle is completed in 2π radians $(360°)$. $\Phi[\ ] = \arctan\left(\dfrac{\text{odd function }[\ ]}{\text{even function }[\ ]}\right).$
χ^2	26	Chi square $\chi^2 = \dfrac{(\chi^2)^{(\nu/2)-1}\exp(-\chi^2/2)}{2^{\nu/2}\Gamma(\nu/2)}$
$\int$	21	Integral.
	24	Estimate.
$!$	35	Factorial, i.e. $n! = n \times (n-1) \times (n-2) \times \ldots \times (n-(n-1)) \times 1$ $(0! = 1)$.

Symbol	Page	Definition

Symbol   **Page**   **Definition**

*   59   Convolution, i.e.
$$X[f] * H[f] = \int_{-\infty}^{\infty} X[f]H[f_1 - f]\,\mathrm{d}f.$$

* superscript   40   Conjugate of the variable. Found by changing the sign of i, i.e. if $c[f] = a[f] - ib[f]$, $c^*[f] = a[f] + ib[f]$.

| | 40   Modulus of the function within the lines.

→   56   Approaches, i.e. $\Delta t \to 0$ means 'as Δt approaches zero'.

∞   17   Infinity.

APPENDIX B

Partial classification of references

The following is a partial classification of the reference list. Since many papers include discussions of the *theory* and most include the *spectrum*, these two possible classification groups have been omitted.

1 Subject

1.1 Astronomy

Aller (1951)
Frankiel and Schwartzchild
(1952)
Lagrange (1772)

Rogerson (1955)
Schuster (1906b)
Strutt (1871)
Uberoi (1955)

1.2 Economics

Bennett *et al.* (1968)
Beveridge (1921, 1922)
Box *et al.* (1967)
Brownlee (1917)
Cunnyngham (1963)
Davis (1941)
Godfrey (1965)
Granger (1966)

Granger and Hatanaka (1964)
Granger and Morgenstern (1963)
Hamermesh (1969)
Harkness (1968, 1969)
Nerlove (1964)
Page (1967)
Quandt (1964)
Rosenblatt (1965)

1.3 Engineering

Houbolt (1961)
Jenkins (1963)

Press and Houbolt (1955)
Press and Tukey (1956)

Note that the study of frequency response is a standard part of almost all engineering degree programmes. Hence many references could be cited, especially from electrical engineering.

1.4 Geology

Agterburg (1967)
Harbaugh and Preston (1965)
Krumbein (1959)
Krumbein and Graybill (1965)
Pincus and Dobrin (1966)

Preston (1966)
Preston and Harbaugh (1965)
Sharma and Geldart (1966)
Stone and Dugundi (1965)

1.5 Geomorphology

Anderson and Koopmans (1963)
Bauer *et al.* (1967)
Bryson and Dutton (1961)
Lee and Kaula (1967)
Peixoto *et al.* (1964)

Preston (1966)
Rodriguez-Iturbe and Yovjevich
(1968)
Speight (1965, 1967)
Stone and Dugundi (1965)

1.6 Geophysics

Aki (1955)
Barber (1954, 1966)
Bartels (1935)
Byerly (1965)
Chapman and Bartels (1940)
Danes and Oncley (1962)
Darby and Davies (1967)
Dean (1958)
Derbyshire (1955, 1956)
Elkins (1951)
Grant (1957)
Hamon and Hannan (1963)

Haubrich (1965)
Haubrich and MacKenzie (1965)
Henderson and Zietz (1949)
Herron (1967)
Hinich and Clay (1968)
MacDonald and Ward (1963)
Mesko (1966)
Neidell (1965, 1966)
Rosenback (1953)
Schuster (1897, 1900)
Ward (1960)

1.7 Hydrology

Kudymov and Sokolov (1962)
Platzman and Rao (1964)
Quimpo (1968a, 1968b)
Rodriguez-Iturbe and Nordin
(1968)

Rodriguez-Iturbe and Yovjevich
(1968)
Yevdjevich and Roesner (1966)

1.8 Meterology and climatology

Angell and Korshover (1963)
Boville and Kwizak (1959)
Brooks and Carruthers (1953)
Brown (1964)
Bryson and Dutton (1961)
Busch and Panofsky (1968)
Buys-Ballot (1847)
Carson (1963)
Chapman (1969)
Charney (1947)
Charnock (1957)
Chiu (1960)
Clayton (1917)
Conrad and Pollack (1950)
Craddock (1957)
Cressman (1958)
Deland (1964)
Eliasen (1958)
Eliasen and Machenhauer (1965)
Estoque (1955)
Everett (1860)
Fitzpatrick (1964)

Forbes (1846)
Gifford (1955)
Gilchrist (1957)
Godson (1959)
Griffith et al. (1956)
Hare (1960)
Harper (1961)
Horn and Bryson (1960)
Julian (1966)
Justus (1967)
Katz (1964)
Katz and Doyle (1964)
Kung and Soong (1969)
Lahey et al. (1958)
Landsberg et al. (1959)
Leese and Epstein (1963)
Lumley and Panofsky (1964)
Oort and Taylor (1969)
Panofsky (1953, 1955, 1967)
Panofsky and Brier (1958)
Panofsky and Deland (1959)
Panofsky and McCormick (1954)

Panofsky and Wolff (1957)
Platzman (1960)
Polowchak and Panofsky (1968)
Priestley (1958)
Rayner (1965, 1967b)
Rosenfeld et al. (1965)
Sabbagh and Bryson (1962)
Saltzman (1957, 1958)
Saltzman and Fleisher (1960)
Saltzman and Teweles (1964)
Schuster (1898)
Schwerdtfeger and Prohaska
                              (1956)

Silberman (1954)
Soong and Kung (1969)
Taylor (1920, 1935, 1938)
Thomson (1860)
Van der Hoven (1957)
Van Loon and Jeune (1969)
Ward and Shapiro (1961)
Webb (1955)
Wiin-Nielsen (1959)
Wiin-Nielsen et al. (1963, 1964)
Winston (1960)
Yen and Dotson (1969)

1.9   Oceanography

Barber (1954)
Cartwright and Rydill (1957)
Cox and Lewis (1966)
Cox and Munk (1954)
Darwin (1883)
Ewing (1969)
Fomin (1968)
Groves (1955)
Hasselman et al. (1963)
Hogben and Lumb (1967)
Monin (1969)
Munk and Bullard (1963)

Munk and Snodgrass (1957)
Munk and Tucker (1959)
N.A.S. (1963)
Phillips (1958)
Pierson (1960)
Pierson and Marks (1952)
Pierson and Moskowitz (1964)
Rushton and Neumann (1957)
Scott (1969)
Seiwell (1949)
Seiwell and Wadsworth (1949)

1.10   Sound

Flanagan (1967)
Maling et al. (1967)
Noll (1964)

Rayleigh (1880)
Singleton and Poulter (1967)

2 Type of analysis
2.1   Cross spectral analysis in one dimension

Akaike (1967)
Bryson and Dutton (1961)
Chiu (1960)
Cooley et al. (1967b)
Estoque (1955)
Goodman (1957)
Granger and Hatanaka (1964)
Hamermesh (1969)
Hamon and Hannan (1963)
Hinich and Clay (1968)

Jenkins (1962, 1963, 1965)
Jenkins and Watts (1968)
Julian (1966)
Panofsky (1955, 1967)
Panofsky and Brier (1958)
Panofsky and Wolff (1957)
Rayner (1967b)
Tick (1962, 1967)
Tukey (1959a, 1959b)

2.2   **Discrete spectral analysis in one dimension**
Angell and Korshover (1963)   Hartley (1949)
Beveridge (1921, 1922)   Horn and Bryson (1960)
Brooks and Carruthers (1953)   Kendall (1945, 1946)
Brownlee (1917)   Lahey et al. (1958)
Chapman and Bartels (1940)   Peixoto et al. (1964)
Conrad and Pollack (1950)   Rayleigh (1880)
Craig (1912)   Sabbagh and Bryson (1962)
Davis (1941)   Saltzman (1958)
Eliasen (1958)   Schuster (1897, 1898, 1900,
Everett (1860)   1906a, 1906b)
Fisher (1929)   Thomson (1860)
Fitzpatrick (1964)   Tolstov (1962)
Forbes (1846)   Van Loon and Jeune (1969)
Freeman (1872)   Walker (1914)
Goertzel (1958)   Whittaker and Robinson (1924)
Hamming (1962)   Wiin-Nielsen et al. (1964)
Harbaugh and Preston (1965)   Yen and Dotson (1969)
Hare (1960)

2.3   **Fast Fourier transform**
Alsop and Nowroozi (1966)   Hinich and Clay (1968)
Bingham et al. (1967)   I.E.E.E. (1967)
Cochran et al. (1967)   Jenkins and Watts (1968)
Cooley (1966)   Jones (1965)
Cooley and Tukey (1965)   Maling et al. (1967)
Cooley et al. (1967a, 1967b)   Oort and Taylor (1969)
Flanagan (1967)   Singleton (1967)
Gentleman and Sande (1966)   Singleton and Poulter (1967)
Helms (1967)   Tukey (1967)

2.4   **Filtering**
Alavi and Jenkins (1965)   Helms (1967)
Bauer et al. (1967)   Holloway (1958)
Byerly (1965)   Krumbein (1959)
Craddock (1957)   MacDonald and Ward (1963)
Cutrona et al. (1960)   Montgomery and Brome (1962)
Dobrin et al. (1965)   Roetling and Hemmil (1962)
Groves (1955)   Tobler (1969a, 1969b)

Note that filtering is involved in the spectral analysis of non-periodic data.   See also references listed under Pattern recognition.

2.5   **Higher-order spectra**
Brillinger (1965)   Hasselman et al. (1963)
Brillinger and Rosenblatt   Rosenblatt (1966)
                (1967a, 1967b)   Tukey (1959a)
Godfrey (1965)

2.6 **Optical analysis**
Bauer *et al.* (1967)
Born and Wolf (1959)
Cheng *et al.* (1968)
Cutrona (1966)
Cutrona *et al.* (1960)
DeVelis and Reynolds (1967)
Dobrin *et al.* (1965)
Elias *et al.* (1952)
Goldstein and Rosenfeld (1964)
Goodman (1968)
Hawkins and Munsay (1963)
Horwitz and Shelton (1961)
Kovásznay and Arman (1957)
McLachlan (1962)

Minot (1959)
Montgomery and Brome (1962)
O'Neill (1956, 1963)
Pincus and Dobrin (1966)
Pollock *et al.* (1963)
Rhodes (1953)
Roetling and Hemmil (1962)
Rosenfeld (1962, 1965)
Rosenfeld and Goldstein (1964)
Rosenfeld and Pfaltz (1966)
Rosenfeld *et al.* (1965)
Stroke (1963)
Tippett *et al.* (1965)

2.7 **Pattern recognition**
Bauer *et al.* (1967)
Cheng *et al.* (1968)
Doyle (1962)
Goldstein and Rosenfeld (1964)
Horwitz and Shelton (1961)
Katz (1964)
Katz and Doyle (1964)

Kovásnay and Arman (1957)
McLachlan (1962)
Pincus and Dobrin (1966)
Pollock *et al.* (1963)
Rosenfeld (1962)
Tippett *et al.* (1965)

2.8 **Two-dimensional analysis**
Bartlett (1963, 1964)
Byerly (1965)
Casetti (1966)
Cox and Munk (1954)
Danes and Oncley (1962)
Darby and Davies (1967)
Dean (1958)
Doyle (1962)
Eliasen and Machenhauer (1965)
Elkins (1951)
Harbaugh and Preston (1965)
Henderson and Zietz (1949)
Katz (1964)
Katz and Doyle (1964)
King (1969)
Krumbein (1959)

Krumbein and Graybill (1965)
Lee and Kaula (1967)
Leese and Epstein (1963)
Longuet-Higgins (1955, 1957)
Matern (1960)
Mesko (1966)
N.A.S. (1963)
Pierson (1960)
Pierson and Marks (1952)
Pierson and Moskowitz (1964)
Preston (1966)
Rosenback (1953)
Stone and Dugundi (1965)
Tobler (1969a, 1969b)
Tolstov (1962)

Also see optical analysis

1.6 Each person's attempt at drawing an interpolation time will be different. Compare with Figure 4.10.4 which has been produced by fitting periodic functions by least squares.

2.6.1

k	0	1	2	3	4	5	6	7	8	9
$A[k]$	7	0	0	2	5	0	0	0	6	0
$\Phi[k[$	0	0	0	30	45	0	0	0	−36	0

Method; match k's with possible terms in Equation (2.4.1) and read off the corresponding magnitudes.

k	0	1	2	3	4	5	6	7	8	9
$a[k]$	4	0	0	0	0	3	0	5	0	0
$b[k]$	0	0	0	0	1	4	0	12	0	0
$A[k]$	4	0	0	0	1	5	0	13	0	0
$b[k]/a[k]$	0	0	0	0	∞	1·33	0	2·4	0	0
$\Phi[k]$	0	0	0	0	90	53° 8′	0	67° 23′	0	0

Method: match k's with possible terms in Equation (2.4.2) and use $a[k]$ and $b[k]$ with Equations (2.3.8) and (2.3.9).

2.6.2

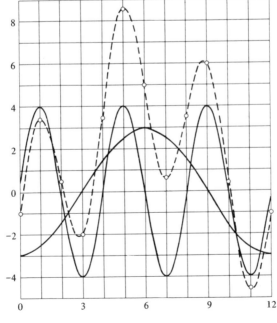

Note that the simplest way to calculate the ordinate points on a sinusoidal curve is to use only $A[k]\cos(k\theta)$ and then move the whole curve to the right or left by $\Phi[k]/k$.

2.6.3 Method: use Equations (2.3.4) and (2.3.5).

k	0	1	3
$a[k]$	2	−3	0
$b[k]$	0	0	4

3.8.1

	P	Q	R	S	T	U	V	W	Y	a	b
0	37		50	24	19	81	−25·5	−19·919	43·5	4·58333	
1	53	−17	67	39	36	98	19·5	33·775	−14·5	12·87917	−5·7365
2	4	−4	−31	39	−40	−62	18	27·713		21·6666	4·61883
3	−31	11	−31		32	2	49	1·732		−2·5	−10·333
4	−35	−19	−51	17						0·1666	0·2886
5	14	−34	−23	15						1·62083	0·90316
6	13		11							−1·4166	

k	A^2	A	Φ first quadrant	$\Phi[k]$	$\Phi[k]/k$
0		4·5833			
1	198·780	14·1	24°	336°	336°
2	490·775	22·2	12° 4′	12° 4′	6° 2′
3	113·027	10·6	76° 22′	256° 22′	85° 27′
4	0·1111	0·3	60°	60°	15°
5	3·4428	1·9	29° 7′	29° 7′	5° 50′
6	2·0069	1·4		180°	30°

3.8.2

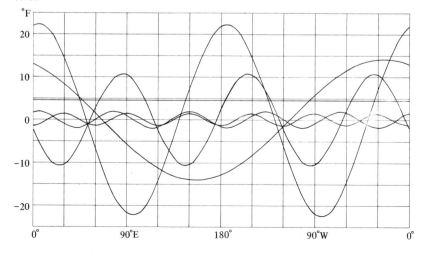

3.8.3

k	$\hat{\sigma}^2$	$\%\hat{\sigma}^2$
0		
1	99·39	24·5
2	245·39	60·6
3	56·52	14·0
4	0·06	0·0
5	1·72	0·4
6	2·01	0·5
	405·09	

3.8.4

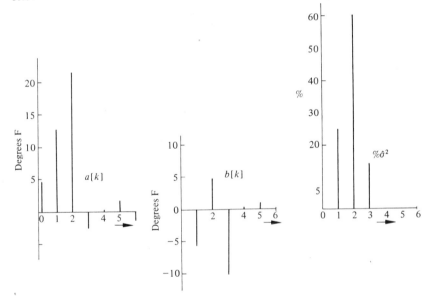

4.10.1 $a[11] = a[1]$ and $b[11] = -b[1]$.

4.10.2

k	$c[k]$	k	$c[k]$	k	$c[k]$
−6	−0·7083	−2	13·1427	3	3·916
−5	1·2619	−1	3·5713	4	−0·0610
−4	0·2276	0	4·5833	5	0·3588
−3	−6·416	1	9·3078	6	0·7083
		2	8·5239		

There are more frequencies than the number of datum points because one of the end points $(-n/2$ or $n/2)$ belongs to a separate periodic set. In order to account for the extra frequency (it appears twice) the magnitude is halved. An alternative would be to calculate the spectrum from $k = 0$ to $k = n$ with $c[6] = 1 \cdot 4166$.

4.10.4

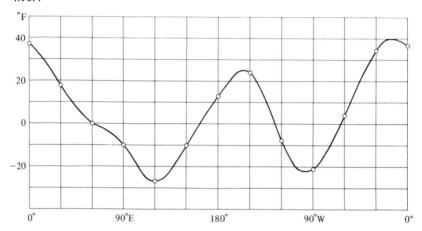

5.5.1 For the one-sided spectrum $a[f] = a[k] \times n\Delta t$. $a[f]$ is the average density in a given band which might fluctuate wildly. A smooth curve is acceptable instead of the histogram below so long as the relationship to $a[f]$ is maintained.

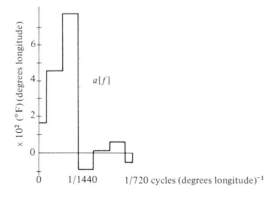

5.5.2 Obviously first differences calculated by the convolution theorem must be the same as those calculated by the direct method.

5.5.3

j	$\dfrac{d(x[t])}{dt}[j]$	j	$\dfrac{d(x[t])}{dt}[j]$	j	$\dfrac{d(x[t])}{dt}[j]$
0	$-0\cdot3810$	4	$-0\cdot0484$	8	$-1\cdot194$
1	$-0\cdot8017$	5	$0\cdot8383$	9	$0\cdot3561$
2	$-0\cdot2774$	6	$0\cdot7437$	10	$0\cdot157$
3	$-0\cdot6382$	7	$-0\cdot3874$	11	$0\cdot6329$

Units are degrees Fahrenheit per degree longitude.

5.5.4

$$\left\{\int x[t]\,dt\right\}[j] = \sum_{k=0}^{n/2} \frac{n}{2\pi k}\left\{-b[k]\cos\left(\frac{2\pi kj}{n}\right)+a[k]\sin\left(\frac{2\pi kj}{n}\right)\right\} + \text{const.}$$

6.5.1 A non-overlapping five-point running mean will reduce the number of observations to 20. Therefore, the number of calculable frequencies are 11 ($k = 0$ to $n/2$) for the one-sided spectrum and 20 for the two-sided spectrum. The response function $W^*[f] = a_w[f]+ib_w[f]$ but, since the function is even, $b_w[f] = 0$. Note that for a response function the density $a_w[f]$ is used and *not* $a_w[k]$. The following densities may be obtained by transforming the weighting function $w[j] = \frac{1}{5}$, $j = 0, 1, 2, 98, 99$; $w[j] = 0, 3 \leqslant j \leqslant 97$. Alternatively, use may be made of the approximate formula for averaging over N terms

$$W^*[f] = \frac{\sin N\pi f}{N\sin \pi f}$$

$k = fn\Delta t$	$a_w[f]$	$k = fn\Delta t$	$a_w[f]$	$k = fn\Delta t$	$a_w[f]$
0	$1\cdot0$	4	$0\cdot938$	8	$0\cdot764$
1	$0\cdot996$	5	$0\cdot904$	9	$0\cdot708$
2	$0\cdot984$	6	$0\cdot836$	10	$0\cdot647$
3	$0\cdot964$	7	$0\cdot816$		

For the two-sided spectrum the effect of the weighting function will be the same at symmetrically matched frequencies, i.e.

$$\text{for } k = 2, f = \frac{2}{n\Delta t}\;; \qquad a[f] = a[-f] = a[f_F - f] = 0\cdot984.$$

This density function should be plotted and compared with Figures 6.3.3.1 and 6.3.3.2. Note that the number of weighting points controls the number of times $W^*[f]$ crosses the zero line.

6.5.2

k	$\Phi[k]$ (deg)	k	$\Phi[k]$ (deg)
0		6	4° 31′
1	44′	7	5° 22′
2	1° 27′	8	6° 16′
3	2° 11′	9	7° 12′
4	2° 55′	10	8° 11′
5	3° 43′		

6.5.3

j	Modified $x[j]$	j	Modified $x[j]$	j	Modified $x[j]$
0	39·13	4	−21·66	8	−3·722
1	27·70	5	−2·605	9	−11·35
2	−0·7783	6	13·37	10	1·157
3	−22·82	7	11·13	11	25·44

6.5.4 Here $W^*[f]$ must be converted to $w[j]$. Again as in Exercise 6.5.1 this step may be performed by direct transformation or, because $W^*[f]$ is even, by the use of the approximate formula given in Section 6.5.1. However, the equivalent of $a_w[k]$ is required rather than the density $a_w[f]$

j	$w[j]$	j	$w[j]$
0	0·4167	6	0·0833
1	0·3110	7	0·0223
2	0·0833	8	−0·0833
3	−0·0833	9	−0·0833
4	−0·0833	10	0·0833
5	0·0223	11	0·3110

7.9.1 In a short series the total variance calculated from alternate observations will usually be different from that calculated from the full series. Furthermore, the variance will be different depending on whether the reduced series starts on the first or second observation. Consequently both are listed below

Series starting with first observation				Series starting with second observation			
k	$\%\hat{\sigma}^2$	k	$\%\hat{\sigma}^2$	k	$\%\hat{\sigma}^2$	k	$\%\hat{\sigma}^2$
0		10	2·0	0		10	3·0
1	2·9	11	4·7	1	6·5	11	7·0
2	1·1	12	1·9	2	1·9	12	17·9
3	10·8	13	4·7	3	1·3	13	0·9
4	1·7	14	0·3	4	9·3	14	1·1
5	7·0	15	9·9	5	5·3	15	0·1
6	12·2	16	2·4	6	14·4	16	6·2
7	7·6	17	2·0	7	5·6	17	2·9
8	2·1	18	10·6	8	4·1	18	8·3
9	0·3	19	15·1	9	4·0	19	0·3
		20	0·5			20	0·0
		Total	0·044777			Total	0·13004

Total for 80 observations 0·088758.

7.9.2 From Equation (7.7.4) we find the following

k	$E = 0·6$	$E = 0·9$	k	$E = 0·6$	$E = 0·9$
0	0·40	0·10	4	1·40	1·65
1	0·57	0·50	5	1·55	1·84
2	0·87	0·95	6	1·60	1·90
3	1·17	1·35			

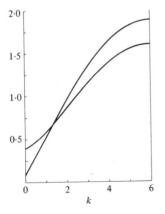

7.9.3

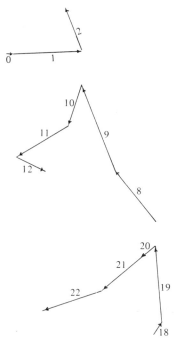

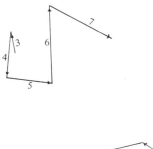

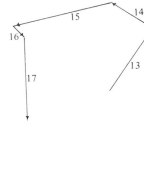

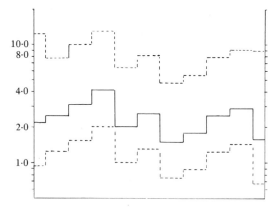

7.9.4 From the information given in the example in Section 7.8
$\nu = 5(80-8)/(100/2)$. From tables, ν/χ^2 [ν, probability level] are
calculated to be

$$\frac{7}{\chi^2[7, 5\%]} = 3 \cdot 23, \quad \frac{7}{\chi^2[7, 95\%]} = 0 \cdot 50, \quad \frac{4}{\chi^2[4, 5\%]} = 5 \cdot 63,$$

$$\frac{4}{\chi^2[4, 95\%]} = 0 \cdot 43.$$

8.7.1 The simple correlation is given by $\sum xy/n\sigma_x\sigma_y$, where x and y are deviations from the mean. Let

$$x[j] = A_x[k]\cos\left(\frac{2\pi jk}{n} - \Phi_x[k]\right)$$

and

$$y[j] = A_y[k]\cos\left(\frac{2\pi jk}{n} - \Phi_y[k]\right)$$

then

$$\frac{\sum xy}{n} = \frac{1}{n}A_x[k]A_y[k]\left\{\sum_{j=0}^{n-1}\cos\left(\frac{2\pi jk}{n} - \Phi_x[k]\right)\cos\left(\frac{2\pi jk}{n} - \Phi_y[k]\right)\right\}$$

but $\cos(A-B)\cos(A-C) = \frac{1}{2}\cos(2A-B-C) + \frac{1}{2}\cos(C-B)$;

$$\sum_{1\ \text{cycle}}\cos A = 0; \text{ and } \sum_0^{n-1}\cos(\text{constant}) = n\cos(\text{constant}).$$

Therefore

$$\frac{xy}{n} = \frac{A_x[k]A_y[k]n}{n}\frac{}{2}\cos(\Phi_y[k] - \Phi_x[k])\ .$$

By Equation (3.4.3) $\sigma_x[k] = A_x/(2)^{\frac{1}{2}}$ and $\sigma_y[k] = A_y/(2)^{\frac{1}{2}}$. Therefore the simple correlation coefficient is

$$\cos(\Phi_y[k] - \Phi_x[k]).$$

8.7.2 The correlation coefficient would be zero. Coherence is given by

$$\hat{R}_{yx}^2[k] = \frac{\{(a_x[k]a_y[k] + b_x[k]b_y[k])/2\}^2}{\{(a_x^2[k]+b_x^2[k])/2\}\{(a_y^2[k]+b_y^2[k])/2}$$

$$+ \frac{\{(a_x[k]b_y[k] - a_y[k]b_x[k])/2\}^2}{\{(a_x^2[k]+b_x^2[k])/2\}\{(a_y^2[k]+b_y^2[k])/2}$$

$$= 1$$

regardless of the magnitudes of the a's and b's. Note that normally the quadrature and variance spectra and the cospectrum are summed over several k values.

8.7.3
Equation (8.5.2.2) gives
    (a) $\tan 1° \pm 0\cdot 1330$;   (b) $\tan 1° \pm 0\cdot 3259$;
    (c) $\tan 89° \pm 436\cdot 61$;   (d) $\tan 89° \pm 1069\cdot 5$.
Equation (8.5.2.3) gives
    (a) $1° \pm 11° 31'$;   (b) $1° \pm 29° 18'$;
    (c) $89° \pm 11° 31'$;   (d) $89° \pm 29° 18'$.

9.7.1 These two-dimensional arrays may be calculated by hand. Note that the possible angles are $0°$, $90°$, $180°$, and $270°$. Therefore the a's and b's result from simple additions and/or subtractions of the data $x[j_1, j_2]$.

(1a)

$a_x[k_1, k_2]/2$

$k_2 \downarrow$ / $k_1 \rightarrow$	0	1	2	3
0	$\frac{65}{16}$	$\frac{3}{16}$	$-\frac{19}{16}$	$\frac{3}{16}$
1	$\frac{1}{16}$	$-\frac{3}{16}$	$\frac{3}{16}$	$\frac{3}{16}$
2	$-\frac{3}{16}$	$\frac{5}{16}$	$-\frac{3}{16}$	$\frac{5}{16}$
3	$\frac{1}{16}$	$\frac{3}{16}$	$\frac{3}{16}$	$-\frac{3}{16}$

$b_x[k_1, k_2]/2$

$k_2 \downarrow$ / $k_1 \rightarrow$	0	1	2	3
0	0	$-\frac{19}{16}$	0	$\frac{10}{16}$
1	$-\frac{8}{16}$	$-\frac{6}{16}$	$\frac{2}{16}$	$\frac{4}{16}$
2	0	0	0	0
3	$\frac{8}{16}$	$-\frac{4}{16}$	$-\frac{2}{16}$	$\frac{6}{16}$

(1b)

$k_2 \downarrow$ / $k_1 \rightarrow$	-2	-1	0	1	2
0			$\frac{65}{16}$	$\frac{6}{16}$	$-\frac{19}{16}$
1	$\frac{3}{16}$	$\frac{6}{16}$	$\frac{2}{16}$	$-\frac{6}{16}$	$\frac{3}{16}$
2	$-\frac{3/2}{16}$	$\frac{5}{16}$	$-\frac{3}{16}$	$\frac{5}{16}$	$-\frac{3/2}{16}$

$k_2 \downarrow$ / $k_1 \rightarrow$	-2	-1	0	1	2
0			0	$\frac{20}{16}$	2
1	$\frac{2}{16}$	$\frac{8}{16}$	$-\frac{16}{16}$	$-\frac{12}{16}$	$\frac{2}{16}$
2	0	0	0	0	0

(2)

$k_2 \downarrow$ / $k_1 \rightarrow$	-2	-1	0	1	2
0	$A_x[k_1, k_2]$		$\frac{65}{16}$	$\frac{\sqrt{436}}{16}$	$\frac{19}{16}$
1	$\frac{\sqrt{13}}{16}$	$\frac{10}{16}$	$\frac{\sqrt{260}}{16}$	$\frac{\sqrt{180}}{16}$	$\frac{\sqrt{13}}{16}$
2	$\frac{3/2}{16}$	$\frac{5}{16}$	$\frac{3}{16}$	$\frac{5}{16}$	$\frac{3/2}{16}$

(3)

$k_2 \downarrow$ / $k_1 \rightarrow$	-2	-1	0	1	2
0	$\sigma_x^2[k_1, k_2]$		0	$\frac{218}{256}$	$\frac{361}{256}$
1	$\frac{13}{256}$	$\frac{50}{256}$	$\frac{130}{256}$	$\frac{90}{256}$	$\frac{13}{256}$
2	$\frac{9/2}{256}$	$\frac{25}{256}$	$\frac{9}{256}$	$\frac{25}{256}$	$\frac{9/2}{256}$

9.7.2

$$\hat{XX}[m_1, m_2] = \sum_{k_2 = n_2/2 - z_2}^{n_2/2 - 1} \left(\sum_{k_1 = n_1/2 - z_1}^{n_1/2 - 1} \frac{\hat{a}_x^2[k_1, k_2] + \hat{b}_x^2[k_1, k_2]}{2} \right.$$

$$\left. + \frac{\hat{a}_x^2[n_1/2, k_2] + \hat{b}_x^2[n_1/2, k_2)}{4} \right)$$

$$+ \sum_{k_1 = n_1/2 - z_1}^{n_1/2 - 1} \frac{\hat{a}_x^2[k_1, n_2/2] + \hat{b}_x^2[k_1, n_2/2]}{4} + \frac{\hat{a}_x^2[n_1/2, n_2/2]}{8} .$$

9.7.3 Since variances are needed for rectangles 4×5 in extent for 0's, the transformed $x^2[j_1, j_2]$ may be multiplied by a transformed rectangle $w[j_1, j_2]$, which is the same size as $x[j_1, j_2]$ with a 4×5 set of $1/20$ths in the upper left corner. The square root of the retransformed result will be the two-dimensional variation of the standard deviation required for $x[j_1, j_2]$. The target standard deviation will be a constant and will not affect the location of the maxima in the correlation array.

9.7.4 The whole average array will be moved along the diagonal, i.e.

4·25	4·0	3·0	3·25	4·5	2·25	2·5	4·75
4·75	4·5	2·25	2·50	5·0	3·5	4·0	5·5
5·5	5·0	3·5	4·0	4·5	4·25	4·75	5·0
5·0	4·5	4·25	4·75	4·0	3·0	3·25	4·25